LAB WORKBOOK

Principles of Residential Construction

Essential Skills and Practices

Peter Kruppenbacher

Leonard Friedman

Publisher
The Goodheart-Willcox Company, Inc.
Tinley Park, IL
www.g-w.com

ISBN 979-8-89448-042-8

2 3 4 5 6 7 8 9 – 26 – 30 29 28 27 26 25

Image Credits. Front cover: Kirkam/Shutterstock.com

Introduction

This lab workbook is designed to be a companion with the text *Principles of Residential Construction* and to build on the concepts introduced and explored there. All activities in the lab workbook correspond to chapters in the text and should be completed after reading the appropriate text chapter.

The workbook contains reviews of each textbook chapter to enhance your understanding of the textbook content. The various types of questions include matching, true or false, and multiple choice. These assessments will help you master the technical knowledge presented in the text.

The chapter reviews are followed by hands-on lab activities designed to help students develop skills and experience as they work their way through the phases of residential construction. Skill-building activities allow students to practice new skills they will likely need while on a jobsite and throughout their chosen construction careers, and the long-term project allows students to build upon those skills while creating a corner unit from design through installing fixtures.

Reading *Principles of Residential Construction* in combination with using this workbook will help you acquire a working knowledge of the phases of residential construction while building a foundation of skills that will assist you throughout your career. Completing the skill-building and project activities will provide plenty of opportunity to apply knowledge and reinforce concepts introduced throughout the text.

Contents

CHAPTER 5

CHAPTER 6

CHAPTER 7

CHAPTER 8

CHAPTER 9

Name ______________________________ Date ______________ Class __________

CHAPTER 1 Preparing for Construction

Activity 1-1: Chapter 1 Review

Answer the following questions after studying the chapter.

Lesson 1.1

1. _____ *True or False?* Renovating a home is a form of residential construction. (1.1-1)
2. _____ *True or False?* Wages in the construction industry have increased at a slower rate than most other industries. (1.1-2)
3. _____ Much of the oldest known evidence of construction in the world consists of structures made of which building material? (1.1-3)
   A. Grass
   B. Mud
   C. Sticks
   D. Stone
4. _____ Some of the earliest evidence of large-scale construction comes from _____. (1.1-3)
   A. Australia
   B. Iceland
   C. Mesopotamia
   D. North America

Lesson 1.2

1. _____ Which of these is an example of a hard skill? (1.2-1)
   A. Communicating effectively
   B. Following directions
   C. Measuring materials accurately
   D. Managing time effectively
2. _____ *True or False?* When an employee practices an unethical behavior, they may experience a short-term benefit. (1.2-2)
3. _____ The process of identifying an issue and finding a solution is called _____. (1.2-3)
   A. critical thinking
   B. problem solving
   C. having a safety mindset
   D. having a growth mindset
4. _____ *True or False?* An engaged construction worker will sit down and work whenever possible. (1.2-3)
5. _____ Learning from mistakes is part of having a _____ mindset. (1.2-3)
   A. growth
   B. disengaged
   C. safety
   D. technical

Lesson 1.3

1. _____ *True or False?* The most effective method for understanding the language of construction is to memorize definitions. (1.3-1)
2. _____ Which document authorizes the construction of a new structure? (1.3-1)
   A. Building permit
   B. Change order
   C. Construction drawing
   D. Site plan
3. _____ *True or False?* Site preparation should begin before the final approval for construction so that the land is ready when the approval comes through. (1.3-2)

4. _____ How many batter boards are built for each wall? (1.3-2)
   A. 1
   B. 2
   C. 3
   D. 4

5. _____ The depression or cavity in which the foundation will be built is the _____. (1.3-2)
   A. astragal
   B. benchmark
   C. building line
   D. excavation

Lesson 1.4

1. _____ *True or False?* The frost line varies from one region to another. (1.4-1)

2. _____ *True or False?* The dressed dimensions are larger than the nominal dimensions. (1.4-2)

3. _____ Framing lumber is commonly available in which length? (1.4-2)
   A. 6′
   B. 9′
   C. 12′
   D. 15′

4. _____ Which type of screw is generally installed by hand with a screwdriver? (1.4-3)
   A. Lag screw
   B. Roofing screw
   C. Sheet metal screw
   D. Wood screw

5. _____ *True or False?* Lag bolts are commonly used in applications that require the screw head to be exposed, such as on door hinges. (1.4-3)

6. _____ *True or False?* Structural screws have more shear strength than other construction screws. (1.4-3)

7. _____ Which type of screw is made of hardened steel and has twin-lead threads that can cut and bind into masonry surfaces? (1.4-3)
   A. Concrete screws
   B. Lag screws
   C. Roofing screws
   D. Wood screws

Lesson 1.5

1. _____ Which statement about tape measures is true? (1.5-1)
   A. They are also known as measuring tapes.
   B. They are available in lengths up to 300′.
   C. The hook on the end is designed to slide slightly.
   D. They must be rewound with a crank after measuring.

2. _____ Carpenter's pencils are designed to be exactly _____. (1.5-1)
   A. 1/4″ thick and 1/4″ wide
   B. 1/4″ thick and 1/2″ wide
   C. 1/2″ thick and 1/2″ wide
   D. 1/2″ thick and 1″ wide

3. _____ Which tool is designed with a flange along one side so that it can be placed against the edge of a board and used to make a 90° mark quickly? (1.5-1)
   A. Plumb bob
   B. Spirit level
   C. Dry line
   D. Speed square

4. _____ *True or False?* If something is level, then it is perfectly vertical. (1.5-1)

5. _____ Which tool is used to cut metal fasteners, such as bolts? (1.5-2)
   A. Drywall saw
   B. Hacksaw
   C. Coping saw
   D. Crosscut saw

6. _____ *True or False?* Battery-operated circular saws are more powerful than corded saws. (1.5-2)

7. _____ *True or False?* The smaller the blade diameter is on a circular saw, the deeper the cut it can produce. (1.5-2)

Name ______________________________

Lesson 1.6

1. _____ *True or False?* Many workplace accidents can be prevented simply by using safer work practices. (1.6-1)
2. _____ *True or False?* A construction employee has the right to request that their employer correct hazards in the workplace. (1.6-2)
3. _____ *True or False?* When working with a chisel, you should hold the workpiece with one hand and operate the chisel with the other hand. (1.6-3)
4. _____ *True or False?* If you need to lower a corded tool to someone working below you, it is safest to lower the tool by its cord. (1.6-3)
5. _____ Which type of tool often uses pneumatic power? (1.6-3)
   A. Jackhammer
   B. Demolition saw
   C. Mortar mixer
   D. Pin gun
6. _____ *True or False?* A pin gun is powerful enough to drive a drive pin into concrete. (1.6-3)

Lesson 1.7

1. _____ The longest marks on a ruler indicate _____. (1.7-1)
   A. half-inches
   B. inches
   C. feet
   D. yards
2. _____ Three feet can be written as _____. (1.7-1)
   A. 3′
   B. 3″
   C. 3°
   D. 3;
3. _____ To convert 96 inches to feet, you should perform which calculation? (1.7-1)
   A. $96 \div 3$
   B. 96×3
   C. $96 \div 12$
   D. 96×12
4. _____ A ruler has _____ sixteenths in a half-inch. (1.7-1)
   A. 4
   B. 6
   C. 8
   D. 10
5. _____ $124 - 39 =$ _____. (1.7-2)
   A. 85
   B. 95
   C. 115
   D. 163
6. _____ $216 \div 3 =$ _____. (1.7-2)
   A. 15
   B. 72
   C. 73
   D. 98
7. _____ *True or False?* When setting up a division problem, the dividend is written inside the division symbol. (1.7-2)

Lesson 1.8

1. _____ *True or False?* Section drawings are enlarged drawings that show the details of how a particular aspect of the building is to be constructed. (1.8-1)
2. _____ Reference symbols are used with _____ to refer to another drawing in the set of plans. (1.8-1)
   A. callouts
   B. lines
   C. projections
   D. views
3. _____ *True or False?* The bottom view of a building is typically included in construction drawings. (1.8-1)
4. _____ The layout of a designated level in the building is shown in the _____ plan. (1.8-2)
   A. elevation
   B. foundation
   C. floor
   D. site
5. _____ *True or False?* Site plans often include topographical features. (1.8-3)

Lesson 1.9

1. _____ Green building aims to _____. (1.9-1)
   A. ensure the building lasts as long as similar traditional buildings
   B. move the main floors of the building underground to reduce energy use
   C. construct the building using only naturally produced materials
   D. lessen the overall impact of the building on the environment

2. _____ *True or False?* Simply adding photovoltaic solar panels to a home is enough to make it green. (1.9-1)

3. _____ *True or False?* One benefit of green building is that it poses fewer risks to builders. (1.9-1)

4. _____ Which of these is an example of a green building code? (1.9-1)
   A. American Standard Building Code
   B. International Building Code
   C. International Energy Conservation Code
   D. International Residential Code

Name ______________________________ Date ______________ Class __________

Activity 1-2: Skill Building—Choosing Proper Personal Protective Equipment (PPE)

Learning Outcomes

After completing this lab activity, you will be able to:

- Explain the *margin of safety* when working with sharp tools.
- Identify personal protective equipment (PPE) commonly used by construction workers.
- Verify the proper fit of various types of PPE.

Introduction

Being safe on the job is more than just following rules and regulations; it is a mindset. You need to be aware of possible safety hazards and what you can do to prevent accidents from happening. Being safe in the workplace is up to you.

Wearing proper personal protective equipment (PPE) is vital to workplace safety. In this activity, you will practice key safety procedures to keep in mind when working. You will also verify a proper fit for various types of PPE. This will help ensure proper adjustments are made to PPE so that they perform their intended function.

Text Reference

Chapter 1, *Preparing for Construction*

Equipment

The following tools and equipment are needed to perform these activities:

- Cell phone or dollar bill
- Hearing protection, such as earmuffs and earplugs
- Hard hat
- Measuring tape
- N95 dust mask and respirator
- Safety boots or closed-toe shoes
- Safety glasses
- Work gloves

Safety Review

Follow all safety procedures specified by your instructor and specified in any appropriate manufacturer's information. The following are some additional safety-related items to keep in mind:

- Store personal protective equipment (PPE) properly.
- Always keep the work area as clean as possible. A clean workspace is a safe workspace.

Procedure

1. Measure the length of a dollar bill or cell phone with the measuring tape. That distance can be referred to as the *margin of safety*, or the distance to maintain between your hands and any sharp tools or blades on machinery. If your hand slips while using a sharp tool, the margin of safety gives you enough space to avoid an injury. Keep the length of a dollar bill or cell phone in mind when working with sharp tools.
2. Put on a pair of safety glasses. Make sure they fit snugly, have the ANSI Z87.1 rating, and are clean.

NOTE

Prescription glasses may be purchased with the ANSI Z87.1 safety rating but must also include removable side shields to meet the standards. Sunglasses may also be purchased with the proper rating but should be used with caution as a substitute for clear safety glasses.

3. Before using a hard hat, verify the method of adjustment in the harness under the helmet. Some hard hats adjust with a turn knob at the rear of the harness, others have a series of holes and pegs that must be pushed together to secure the position of the harness. Put on the hard hat, determine how the harness needs to be adjusted, and tighten or loosen to fit your head. The harness should be snug around your head and fully down so the harness rests on the top of your head. To check the fit, place the helmet on and bend at your hips to touch the floor. While touching the floor, move your head from side to side and up and down making sure the helmet stays in place during this movement. When it stays in place, the adjustment is correct.
4. Check the fit of your clothing. Loose clothes, jewelry, or hair can get caught in moving machinery or power tools. If your shirt is loose, tuck it in. Button or roll up sleeves if applicable. Tie back long hair and remove or tuck in long or dangling jewelry.
5. Put on safety boots or closed-toed shoes. Check for proper traction. If the environment is often wet or slippery, consider non-slip shoes.
6. A dust mask is used to protect from breathing in different types of dust or particulates that become airborne. Put on the dust mask and adjust it to fit your face with the following steps. One elastic strap should go under your ears and around the lower part of your neck. The other strap should go above your ears and over the back of your head. Mold the metal nose piece to follow the shape of your nose and onto your face tightly fitting your shape. Ensure the mask forms a proper seal over your mouth and nose by breathing heavily and feeling the air build up in the mask. If no pressure is built up, too much air is leaking around the mask and adjustments must be made.
7. A respirator is used to protect against harmful fumes and gases released into the environment while working. There should be two straps holding the respirator in place. One should be low on the back of the neck and the other higher on the back of the head. Adjust the fit by sliding the buckle in or out of the strapping to lengthen or shorten the straps. Verify a tight fit again by breathing heavily and feeling the pressure build up inside the mask. If no pressure builds up, tighten the strapping and adjust the position of the mask on your face for a better fit. Different cartridges are available for different types of hazards, ensure the proper cartridge is being used for the hazard present.

NOTE

OSHA requires any respirator or filtering facepiece to be NIOSH approved. A filtering facepiece is a dust mask, but a nuisance dust mask easily obtained at the local hardware store is not NIOSH approved and should not be used on any jobsite. A dust mask with an N95 label is the minimum style dust mask allowed by OSHA standards.

8. Put on the earmuffs. Ensure the padding around the ear cover comes in complete contact with the area around your ears. The band of the earmuffs should rest on the top of your head to help keep them in place. Test them by having a partner talk to you while wearing them. You should still be able to hear your partner, but they should sound muffled. If you can hear them clearly, adjust the fit of the hearing protection or choose a different type of protection. If anything keeps the padding from being in contact with the area around the ear, such as required safety glasses, the protection will be minimal.
9. Foam earplugs offer the best protection from sound when used properly. Roll the earplug into a cylinder as small as possible. With the earplug still rolled in one hand, reach around your head with your other hand and pull your ear back to open it up a bit. Insert the earplug into your ear as far as possible. Let go of your ear, place a finger on the earplug, and keep pressure on it while it expands inside your ear canal. The foam plug will expand and fit your ear canal perfectly providing the best protection. Repeat the same steps for the other ear to prepare to work in a loud environment.

NOTE

The OSHA standard instructs that hearing protection should be worn when a worker is exposed to sound levels above 85 decibels. To measure this, there are many free sound meter apps available for a phone. Download one of these sound meter apps and check what level of sound is in the space you are working in. You might be surprised!

Name ______________________________

10. Put on a pair of work gloves that fit your hands. These are designed to protect your hands when moving materials or working with hazardous chemicals. Test their fit by performing measurements with the measuring tape. The gloves should fit comfortably and not inhibit your ability to take measurements.

NOTE

Gloves should *not* be worn while operating power tools or machinery because they can get caught and cause injuries.

Review and Analysis

1. Why is it important to maintain the *margin of safety*?

____

____

____

2. In what situations would you use a dust mask? In what situations would you use a respirator?

____

____

____

3. The construction industry offers many career paths, including roofer, drywall installer, finish carpenter, asbestos remover, and more. What are some unique PPE requirements in these different careers?

____

____

____

Notes

Name ______________________________ Date ______________ Class __________

Activity 1-3: Skill Building—Measuring and Marking Layout Lines

Learning Outcomes

After completing this lab activity, you will be able to:

- Interpret the markings on a tape measure with precision.
- Mark layout measurements accurately on framing members.
- Utilize layout tools to prepare for cuts or construction processes.

Introduction

Much of what a carpenter does involves measuring and laying out different parts of a project. In construction, workers must be proficient in reading tape measures and using the typical layout tools carried in the tool apron or readily available on the jobsite. This activity will provide practice in interpreting the graduated marks on a tape measure for many layout steps.

When framing a residential structure, there may be some room for slight measurement inaccuracies since the work will be covered by other finish materials. However, when doing the finer finish work of installing cabinets and trim, cuts and installations must be done as precisely as possible. Being able to quickly read and transfer measurements becomes an important skill of the carpenter.

Text Reference

Chapter 1, *Preparing for Construction*

Equipment

The following tools and supplies will be needed for this exercise.

- 1″ × 4″ × 12″ piece of pine
- 2″ × 4″ × 24″ piece of lumber
- Combination Square
- Framing square
- Safety glasses
- Sharp pencil
- Speed square
- Tape measure

Safety Review

Follow all safety procedures specified by your instructor and any appropriate safety rules from the manufacturer. The following are some additional safety-related steps to keep in mind:

- Wear safety glasses at all times
- When handling framing lumber, always be careful of broken edges and splinters.
- Be careful handling the framing square; if dropped, sharp edges can be created, and it can cause the square to become inaccurate.

Procedure

1. On the piece of 1×4×12, use a pencil to make layout marks at the measurements shown in **Figure 1**. Then, use a combination square to draw the lines completely across the board. Mark and draw all the lines across the board and then mark and draw all the lines down the length of the board.

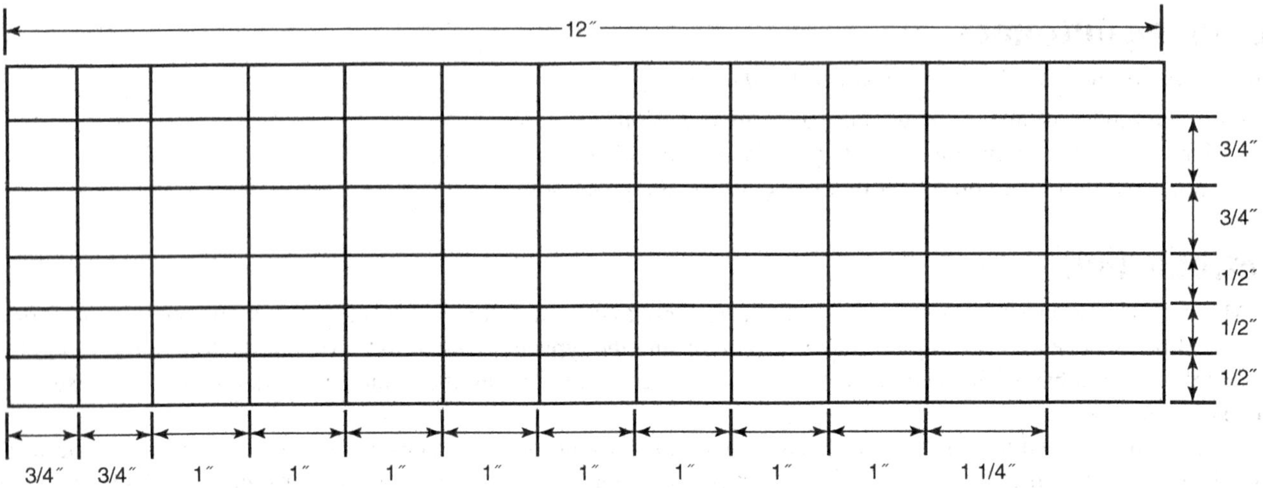

Goodheart-Willcox Publisher

Figure 1

2. Use the combination square to continue the measurement lines from one side of the board to the other. To check the accuracy of the layouts, put the square on the other edge or end of the board and verify that the layout marks line up. If done carefully, the lines will be continuous around the board in each direction, **Figure 2.**

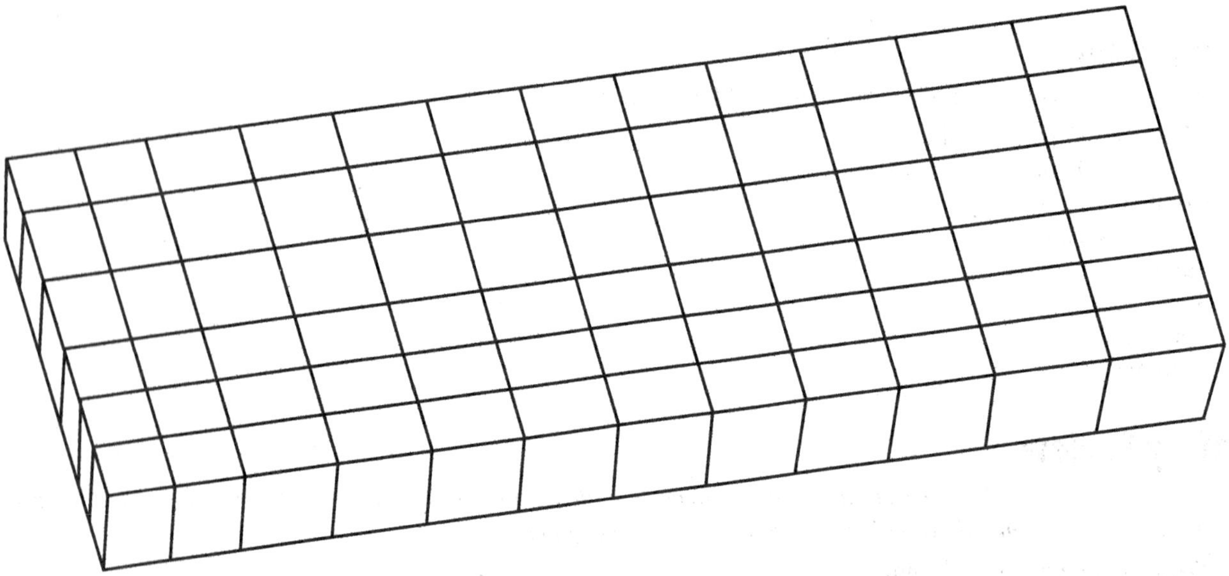
Goodheart-Willcox Publisher

Figure 2

3. Use the tape measure and speed square along the length of the 2×4×24 to lay out the following measurements:

2″	7 5/8″	15 7/16″	20 3/8″
3 1/2″	9 3/16″	16 15/16″	21 9/16″
4 1/4″	11 7/8″	18 1/16″	22 5/16″
6 1/8″	13 11/16″	19 3/4″	

Name ____

Use the crow's foot mark at each measurement to indicate where to draw the line. Then, use the speed square to carefully draw a line across the width of the board that aligns with the crow's foot mark, **Figure 3**. When complete, ask the instructor to check for accuracy.

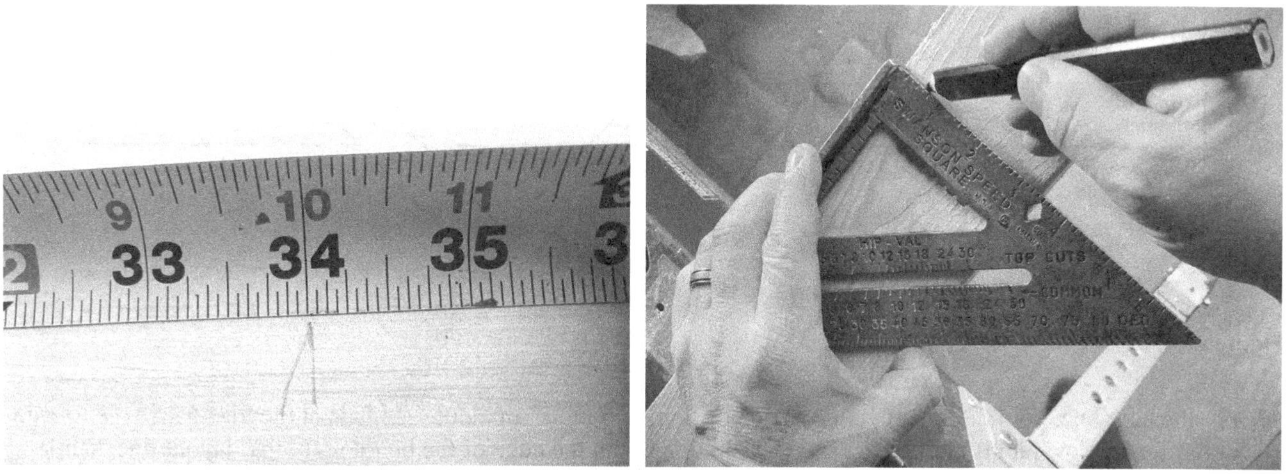

Goodheart-Willcox Publisher

Figure 3

4. On the other side of the 2×4, draw a line across the width of the board at 16″ from the left end of the board. Using the speed square, draw a line across the board at 45° to the left of the line. Start at the bottom edge of the board and intersect the line at 16″.

5. At the base of the board at the same 16″ line, set the speed square against the edge of the board at the pivot point. See Figure 1.5-11B in the textbook for help if needed. Turn the square to line up the 30° mark with the edge of the board and draw a line across the board at that angle to the left of the line as well. Refer to **Figure 4** to check your work.

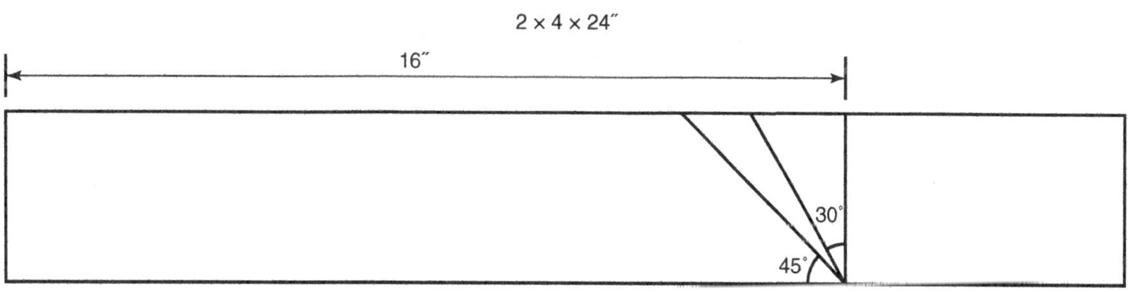

Goodheart-Willcox Publisher

Figure 4

6. Use the framing square to draw the next two lines. Hold the square with the blade of the square in the right hand and the tongue with the left hand. Using the measurements along the outside edge of the square, hold the blade at 7″ along the bottom edge of the board at the 16″ layout line. Carefully line up the 7″ mark on the tongue of the square with the edge of the board to the left while keeping the 7″ at the line on the blade as well, **Figure 5**. When both edges line up with the 7″ mark on each side of the square, draw a line along the blade of the square across the board to the right of the line.

NOTE

When referring to the framing square, the wider longer side of the square is called the blade of the square. The thinner, shorter end of the square is called the tongue of the square. There are measurement marks along each of the edges of the square. Some of these inch marks are broken down into 1/8s of an inch, others are to the 1/16″. Some are divided into tenths of an inch as used by engineers. Be careful when measuring that the correct edge is being used to do the layouts.

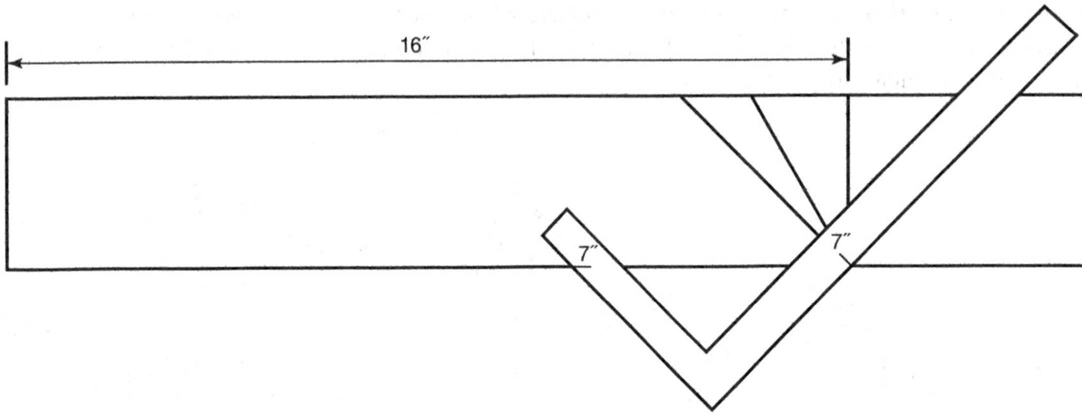

Goodheart-Willcox Publisher

Figure 5

7. For the next line, hold the square in the same position with the 7″ mark of the blade at the line. Move the tongue of the square down until the 12″ measurement lines up with the edge of the board at the 16″ layout line. While holding 7″ on the blade and 12″ on the tongue, draw a line across the board to the right of the blade.
8. Use the speed square to measure the angles that were drawn with the framing square. Line up the pivot point to the bottom of the board at the square line, line up the right edge of the square with each line, and then look at the angle scale to determine each angle. If done correctly, these lines should be the same angles as the other side drawn with the speed square; one is a 45° and the other is a 30° angle. With a little research, angle tables can be found giving numbers to hold on the body and tongue to lay out any angle using the framing square.

Review and Analysis

1. Why are accurate measurements so important in the construction industry?

____

____

____

2. Did the lines on the 1×4×12 line up accurately around all surfaces of the board? If not, what could have been done to make the lines more accurate?

____

____

____

Name ______________________ Date ____________ Class __________

Activity 1-4: Skill Building—Fastening Pieces of Lumber Using Screws

Learning Outcomes

After completing this lab activity, you will be able to:

- Identify the different holes drilled to prepare for the screw.
- Fasten a screw for the best holding and clamping strength.

Introduction

Screws are used through all phases of construction. If the screw is used to pull the lumber together for a tight joint, the screw must be installed correctly. Before installation, screws require a drilled hole to allow the screw to properly grip both pieces without splitting the wood being fastened. A construction worker needs to understand when the proper holes need to be drilled and how to choose the right size hole.

Text Reference

Chapter 1, *Preparing for Construction*

Equipment

The following tools and equipment are needed to perform this activity:

- 1×4×12″ boards
- 2×4×12″ boards
- Cordless drill or driver
- Countersink bit
- Deck or framing screws, 2″
- Drywall screws, 2″
- Safety glasses
- Twist bits, set
- Wood plugs, 3/8″
- Wood screws, 2″

Safety Review

Follow all safety procedures specified by your instructor and specified in any appropriate manufacturer's information. The following are some additional safety-related items to keep in mind:

- Wear safety glasses at all times.
- Be aware of sharp points on the ends of the screws and handle carefully.
- Keep hands away from the drill bit while making the holes.
- Be mindful of loose clothing to avoid getting caught by the drill when working.
- Do not touch the bit after drilling since it will likely be hot.

Procedure

No Holes

1. Align one edge of the 1×4 with one of the edges of the 2×4, creating an L-shape with the two pieces, **Figure 1**.

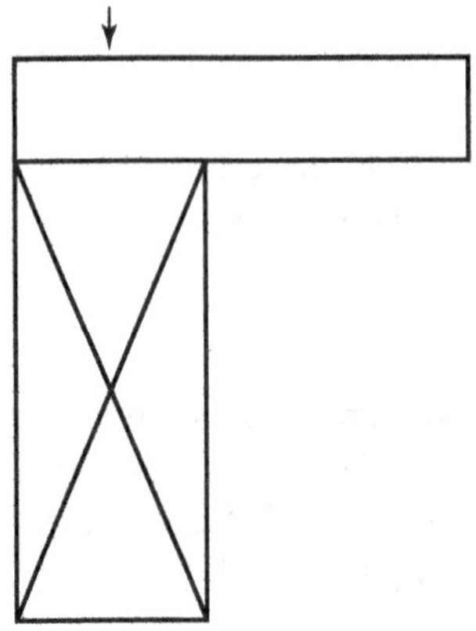

Goodheart-Willcox Publisher

Figure 1

2. While holding the boards together with hand pressure, connect the two pieces by driving a drywall screw into the joint. Stop the screw when the head is flush with the surface of the 1×4 board.
3. Examine the joint and try spinning the 1×4 board on the screw. There is probably a small space between the two pieces, and the piece turns easily. The screw did not create a tight hold. This is what typically happens without the correct holes drilled.
4. Remove the screw after examining the pieces to reuse materials.

Shank Hole

For a screw to pull two pieces of wood tightly together, the screw threads must not grab the first piece of wood. To achieve this, a shank hole, at least as wide as the threads on the screw, is drilled in the first piece of wood.

1. To drill a shank hole, choose a twist bit that is the same size or slightly larger than the width of the threads on the drywall screw, **Figure 2**.

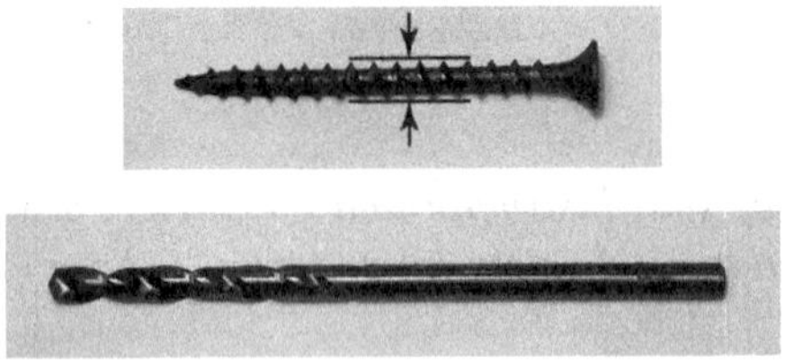

Goodheart-Willcox Publisher

Figure 2

2. On the 1×4 board, use the twist bit to drill a hole through the board 3/4″ from the edge.
3. Place the 1×4 piece on top of the 2×4 piece in the same L-shape and fasten them together using the drywall screw. Now when you try to spin the 1×4, it is considerably harder.
4. Remove the screw after examining the pieces to reuse materials.

Pilot Hole

A pilot hole is another type of hole that helps to secure screws. It allows the screw to enter the second board without splitting the wood. The pilot hole must match the thickness of the metal inside the threaded area of the screw. If drilling into softwood, the pilot hole can be a little smaller to allow the threads to have a tighter grip. A pilot hole should be drilled to a depth equal to the length of the screw.

Name ______________________________

1. Examine the threaded section of the wood screw, **Figure 3**, and choose a drill bit for the pilot hole.

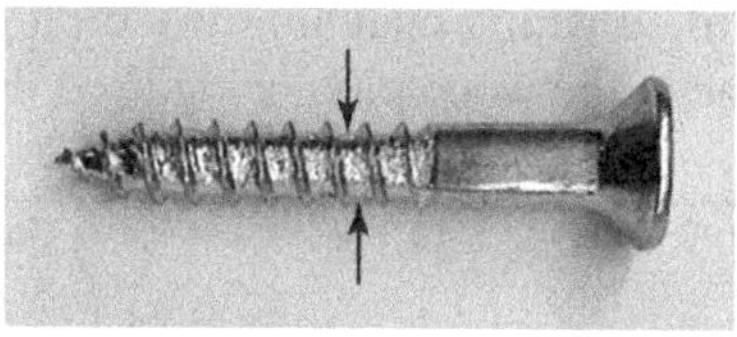

Goodheart-Willcox Publisher

Figure 3

2. With a wood screw, both a pilot hole and a shank hole are needed. On a wood screw, the smooth portion of the screw near the head is the shank size to match when choosing the drill bit for the shank hole, **Figure 4**. You must match the shank size for a wood screw to ensure you do not split the wood.

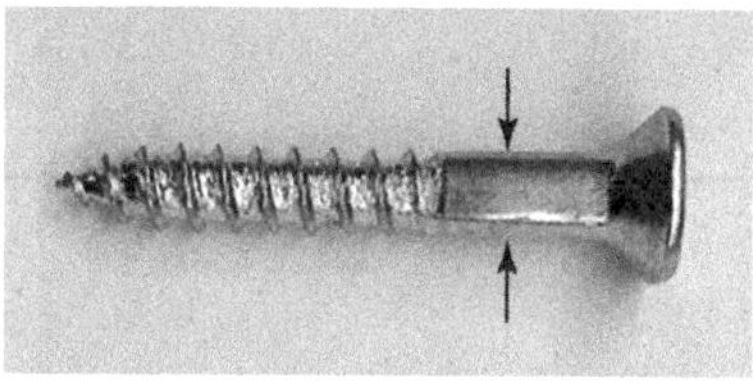

Goodheart-Willcox Publisher

Figure 4

3. Hold the L-shape with the two boards again, aligning the edge of the 1×4 piece along the edge of the 2×4 piece. About 3/4″ from the edge of the 1×4, drill the pilot hole to the depth of the screw using the drill chosen in Step 1.
4. Move the 1×4 to the edge of the bench and drill through the same hole again using the drill chosen in Step 2 to match the shank size. The wood screw should slide through this hole easily.
5. Push the wood screw through the shank hole in the 1×4 piece and use it to line up the pieces into the L-shape by starting the point into the pilot hole in the 2×4. Fasten the two pieces together. Notice how tight this connection becomes with the proper size holes drilled.

Countersink Hole

If the head of a screw needs to be flush with the finished surface, a countersink is used to shape the hole so the head fits easily and goes to the correct depth. This may be needed when working with metals or hardwoods.

1. Drill a new pilot hole through the 1×4 piece of wood to the full depth of the screw.
2. Drill the shank hole ensuring the depth of this hole does not go past the thickness of the 1×4 piece of wood.
3. Use a countersink bit to shape the top of the shank hole to match the size of the head of the screw, **Figure 5**.

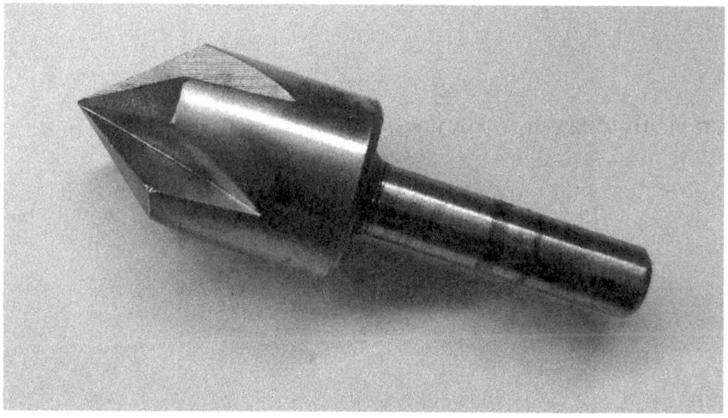

Goodheart-Willcox Publisher

Figure 5

4. Drive a wood screw in this hole.

Counterbore Hole

Sometimes the hole of the screw will be filled with a matching wood plug to hide the screw. To accomplish this, a counterbore hole is drilled to match the size of the wood plug being used, **Figure 6**.

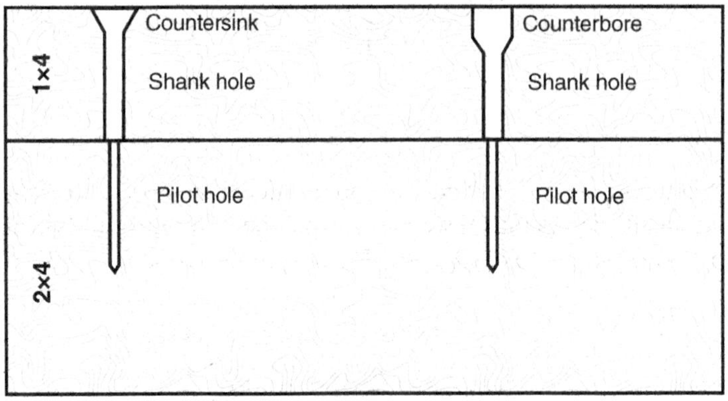

Goodheart-Willcox Publisher

Figure 6

1. In the 1×4 on top of the L-shape, drill the counterbore hole with a 3/8″ bit. Be sure to control the depth of the drill and stop at a depth of 3/8″.
2. Drill the pilot hole to the proper depth.
3. Drill the shank hole only through the top piece of wood.
4. Fasten the boards together with a wood screw.
5. Tap a wood plug in until it is flush with the surface around it.

Self-Drilling Screws

Some screws feature a notch out of the point that creates its own drilling action. With these features, no pilot or shank hole is needed for this screw. It is designed to be easily driven in without these holes to help the construction worker work faster.

1. Examine the deck/framing screw. Notice some differences with this screw compared to the others. The shank of the screw is about the same size as the center of the thread and it has the notch at the point, **Figure 7**.

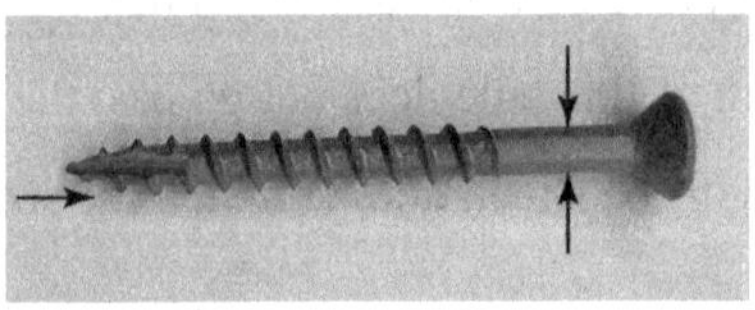

Goodheart-Willcox Publisher

Figure 7

2. Drive a couple of these screws in to see how they work.

Name ________________________________

Review and Analysis

1. How should the pilot hole be changed when the wood being fastened is a harder material?

____

____

____

2. What are some other ways that a screw might be hidden for a better appearance?

____

____

____

3. When driving the deck/framing screw in, how did it feel differently than the wood screw?

____

____

____

Notes

Name ______________________ Date ______________ Class __________

Activity 1-5: Project—Building a Sawhorse

Learning Outcomes

After completing this lab activity, you will be able to:

- Read construction drawings to build a sawhorse.
- Transfer angles using layout tools.
- Use the proper tools needed to construct a sawhorse.

Introduction

A sawhorse is a tool used by the carpenter to support material being worked on and to keep the work off the ground. Sawhorses are generally used in pairs especially when working with long material. They are designed to be portable and stackable. In this activity, you will use a variety of hand and power tools to build a sawhorse. A good rule to remember on any building project is to "measure twice, cut once."

Text Reference

Chapter 1, *Preparing for Construction*

Equipment

The following tools and equipment are needed to perform these activities:

- 1/2″ plywood or equivalent, 12″ × 30″
- 1×4 lumber, 6′ long
- 2×4 lumber, 10′ long
- 2×6 lumber, 36″ long
- Block plane
- Builder's level, 24″ or larger
- Carpenter's pencil
- C-clamps or F-clamps
- Circular saw
- Combination square or speed square
- Construction screws (12), 3″
- Construction screws (24), 1 1/2″
- Countersink bit to match construction screws
- Crosscut saw
- Drill bit, sized 3/32″
- Electric drill with Phillips screwdriver bit
- Phillips screwdriver, No. 2
- Power miter saw
- Safety glasses
- Scratch awl
- Sliding T-bevel
- Table saw with fence and push block
- Wood chisel, 3/4″ or 1″ wide

Safety Review

Follow all safety procedures specified by your instructor and specified in any appropriate manufacturer's information. The following are some additional safety-related items to keep in mind:

- Wear safety glasses at all times.
- When using a chisel, push the chisel away from your body.
- Clamp workpieces down whenever possible.
- Maintain a margin of safety of at least 6″ when working with sharp tools or machinery.

Procedure

For this procedure, follow each step carefully. Refer to the finished sawhorse, **Figure 1**, and construction drawings, **Figure 2**, as you work.

Goodheart-Willcox Publisher

Figure 1

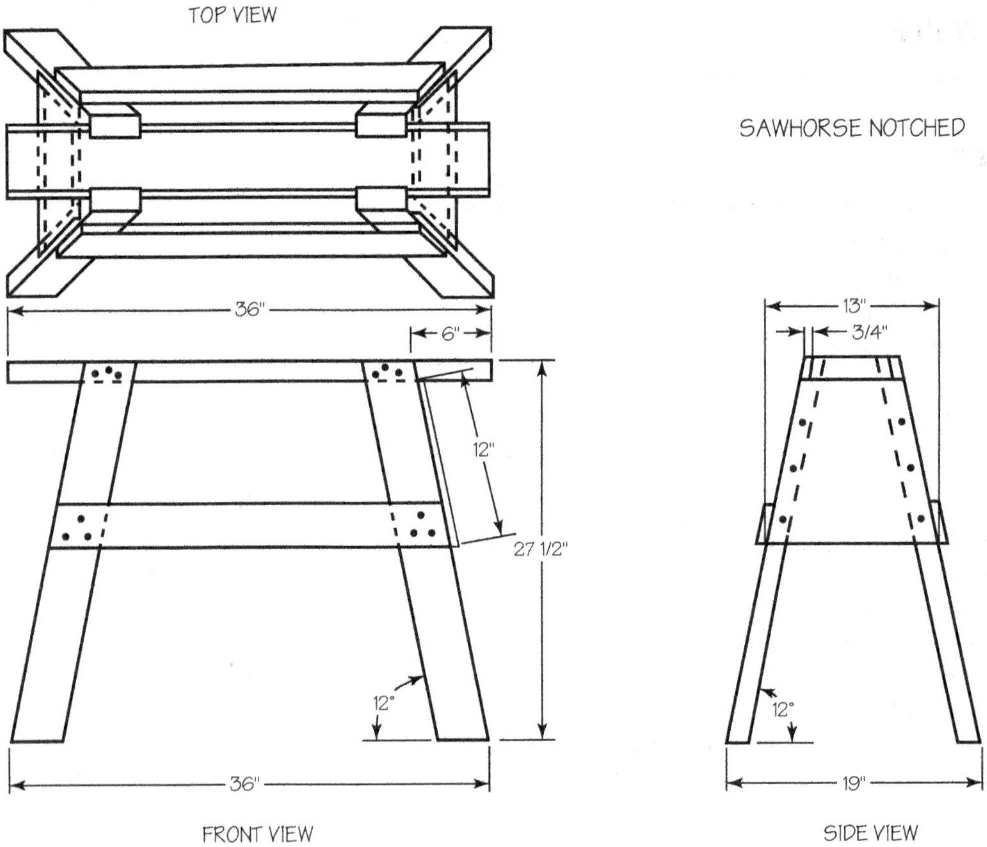

Goodheart-Willcox Publisher

Figure 2

Name ______________________________

Top

1. Use the circular saw or power miter saw to crosscut the 2×6 lumber to 36″ long.
2. Using the table saw, set the fence to 5 1/2″ and rip both edges to 12°. Refer to the side view in **Figure 3**.

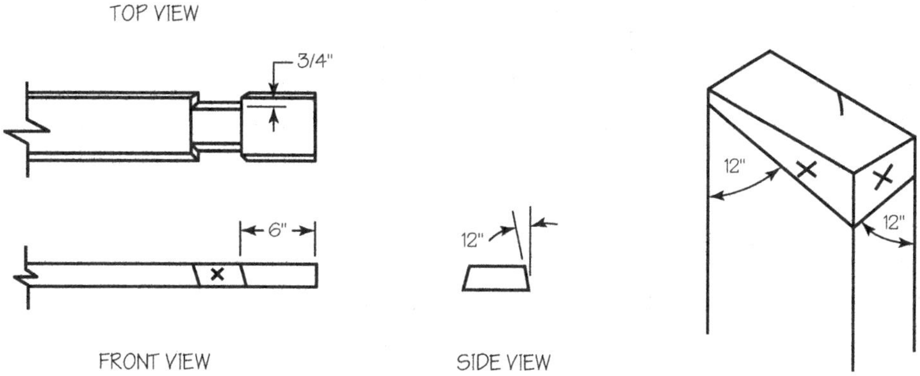

Goodheart-Willcox Publisher

Figure 3

3. Measure 6″ from each end of the 2×6 and use a combination or speed square to draw a square line across the face of the board.
4. Measure the width of the 2×4 that you plan to use for the sawhorse leg and mark that distance from the square line you drew in Step 3 toward the center of the board. Draw the second set of square lines across the face of the board. This layout allows for an exact fit for the legs of the sawhorse.
5. Use the sliding T-bevel to transfer the angle of the 12° cut onto the 2×6. Draw this 12° angle on the edges of the 2×6 so they line up with the square lines you drew on the face of the board, with each leg's angled lines facing toward each end of the board (outwards). Refer to the front view of the sawhorse plans in **Figure 1**.
6. Draw an *X* on the edge of the board where the leg will be notched into.
7. On the face of the board, at both the top and bottom, measure in 3/4″ and mark with an *X*.
8. Place your board in a vise. Use the crosscut saw to cut saw kerfs (cuts in the wood) to be 1/4″ apart the length of the area marked for the first leg, **Figure 4**. These cuts will serve as notches for the first leg to fit into.

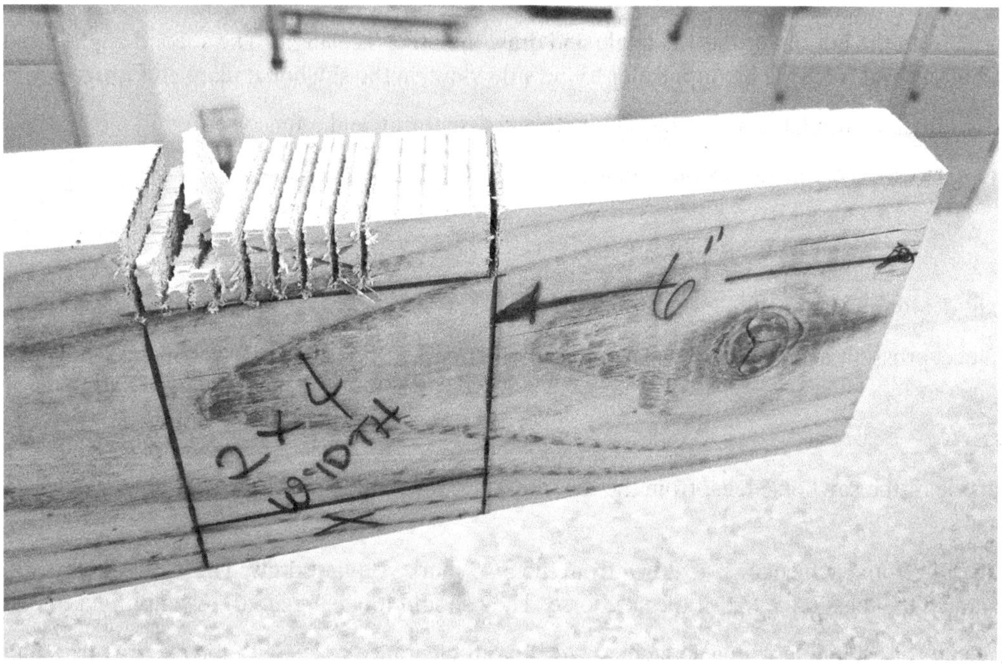

Goodheart-Willcox Publisher

Figure 4

9. With the board still in the vise, carefully chisel the material out in a downward motion. To start, hold the flat side of the chisel parallel to the layout lines and chip out all the notches from the saw. To complete the bottom of the cutout, turn the chisel over so the bevel is down and use the angle of the chisel and the bevel to control the depth of cut. Create a flat surface along the whole bottom of the cutout area. Chisel away from your body and do not cut or chisel the layout lines away.
   - If your notch chips or tears, finish cleaning out the notch with a rasp.
   - Check the fit of the leg to the notch. It should be a press fit and be at a 12° angle. Adjust as necessary. If the notch is a little small, take a hand plane and plane the edge of the leg to fit.
10. Repeat Steps 8–9 to cut notches for the remaining legs. Make sure the angles for the legs face outward.

Legs

11. Crosscut the 2×4 lumber to 30″ long to make four legs.
12. Draw the 12° angle on the face and edge at one end of the first leg.
13. Put this 2×4 leg in a vise or clamp to a bench and cut your 12° compound angle using a crosscut saw. Do *not* cut your marked line away.
14. Copy this same 12° compound angle 29″ down at the bottom of the leg. This angle *must be parallel* with the compound angle at the top of the leg. Refer to the front view on the sawhorse plans in **Figure 2**.
15. Crosscut the 12° compound angle for the bottom of the leg.
16. Cut the next leg using a circular saw. Make the same layout marks as done on the first leg. With the saw unplugged, tilt the saw to a 12° angle and lock it in place. Clamp the leg to a bench or another sawhorse and make the crosscut following the line across the face of the board. Check the direction of the tilt of the blade and be sure it matches the direction of the 12° angle on the edge of the board. Make the crosscut in the direction that matches this angle.
17. Cut the final two legs with the power miter saw. Set the tilt of the blade to 12° and set the angle of the cut to 12° from the fence. Make the first cut on one end of the remaining legs. Then slide the board along the fence to line up the cut with the 29″ mark. Do *not* turn it around or flip it in any direction. You should switch hands making the cut and holding the board since the majority of the board will now be on the other side of the blade.

Gusset

18. Draw a 12° angle on the short side of the 1/2″ plywood for the gusset.
19. Measure 13″ from the bottom of the 12° angle and draw the other 12° angle. The second angle should measure 13″ at the bottom and 5 1/2″ at the top. Refer to the side view on the sawhorse plans in **Figure 2**.
20. Cut this angle with a circular saw or table saw with a crosscut tray and a fence set for 12°.
21. Repeat Steps 18–20 to create the second gusset.

Rail

22. Crosscut the 1×4 lumber in half.
23. On each piece, crosscut a 12° angle on one end. The other end will be cut to 12° during assembly.

Assembly

24. On the outside of the sawhorse legs, from the top, measure down 3/4″. Mark the middle of the legs 1/4″ up from the 3/4″ mark.
25. Measure in 3/4″ from each edge, 1/4″ down from the 3/4″ mark you first drew. These are the layout marks for the screws to attach the legs to the top of the sawhorse. They should look staggered, resembling a triangle.
26. Mark the center of each hole with a scratch awl, drill your clearance holes with the electric drill, and then countersink the holes.

Name ____

27. Attach each leg with **one** 3″ construction screw in the middle hole. Leave the screw a little loose.

NOTE

Fasten all the screws neatly and evenly spaced. As a general rule, screws should be twice the length of the material being fastened.

28. Clamp the top edge of the gusset to the legs, flush with the bottom face of the board that serves as the top of the sawhorse. Next, mark the 3″ construction screw holes with a scratch awl, and countersink if needed. Make sure the gusset is centered before marking and fastening in the screws. Measure down 3″ from the first mark and fasten the first screw to the center of the leg. Repeat the 3″ spacing for the second and third screws. The gusset should now hold the legs evenly spaced.
29. Attach the rail with the 12° cut end flush to the leg on one end and at the same height of the bottom of the gusset. Use 1 1/2″ construction screws, following the same fastening details used in step 23. Make sure both legs are at 12°. Clamp the rail to the other leg. Adjust the legs to make sure both legs are at 12°. Once this is done, mark the rail for length and crosscut your other 12° angle. First unscrew the rail, crosscut at 12°, and then fasten it back on.
30. Tighten the top screws at the top of the legs and add the remaining 3″ screws.

NOTE

If you damage the head of a screw, take it out and replace it with a good screw.

31. Support the sawhorse so you can plane the top flat with a block plane if the legs are protruding above the top of the sawhorse. Plane from the end towards the center so the cut is going with the grain angle. If the cut was made in the other direction, it would be against the grain and would be difficult and cause chip out in the cut.
32. Put the builder's level on the sawhorse and check for level at both the length and width of the sawhorse. Adjust by putting wedges under the legs so the sawhorse sits level.
33. Trim the bottom of the legs until the sawhorse lies flat. One method that can be used to determine how much to trim off the legs is to stand the sawhorse on a flat work surface. Then, place a pencil so it is level with the table surface and trace around the legs with the wedges in place. The tracing serves as a guide when cutting the legs to complete the leveling process.
34. Check for level again.

Review and Analysis

1. How do you determine what length of screw needed to fasten material together?

____

____

____

2. Create a lumber list for a sawhorse that is intended to be 48″ long and 24″ in height.

____

____

____

3. How can a sawhorse be used in a workspace or jobsite?

____

____

____

Notes

Name ______________________ Date ____________ Class ____________

CHAPTER 2 Excavation and Foundations

Activity 2-1: Chapter 2 Review

Answer the following questions after studying the chapter.

Lesson 2.1

1. _____ *True or False?* The job of construction laborer requires specialized or advanced levels of training. (2.1-1)

2. _____ Which member of the construction team is responsible for excavation and moving building materials around the jobsite? (2.1-2)
   A. Mason
   B. Carpenter
   C. Operating engineer
   D. HVAC technician

3. _____ A _____ carpenter installs fine cabinetry and puts the finishing touches on a job. (2.1-2)
   A. cladding
   B. framing
   C. masonry
   D. trim

4. _____ Traditionally, electricians learn the trade by completing an apprenticeship or through _____. (2.1-2)
   A. a technical college
   B. a bachelor's degree program
   C. a master's degree program
   D. another trade, such as plumbing

5. _____ *True or False?* Plumbers working in residential construction must be able to read and follow prints. (2.1-2)

6. _____ *True or False?* Trades workers report directly to the superintendent. (2.1-2)

Lesson 2.2

1. _____ *True or False?* Renovations are a type of residential construction. (2.2-1)

2. _____ *True or False?* Each production home in a development is unique and visually distinct from the other homes within the development. (2.2-2)

3. _____ Which type of home is designed and built to meet the specific needs of an individual client? (2.2-2)
   A. Spec home
   B. Custom home
   C. Semi-custom home
   D. Production home

4. _____ *True or False?* A custom home builder hires many subcontractors to complete different aspects of the home. (2.2-2)

5. _____ Who arranges the capital for a spec home? (2.2-2)
   A. Builder
   B. Buyer
   C. Design firm
   D. Subcontractor

Lesson 2.3

1. _____ *True or False?* When building a home in a cold climate, it is required to locate the foundation above the frost line. (2.3-1)

2. _____ Walls are only a few feet high on a _____ foundation. (2.3-1)
   A. crawl space
   B. ground-supported
   C. monolith slab-on-grade
   D. stem wall

3. _____ *True or False?* A monolithic slab-on-grade foundation serves as both the floor structure and the foundation of the building. (2.3-1)

4. _____ *True or False?* When using a grade beam foundation on soft soil, concrete can be poured directly into the holes for the piers. (2.3-1)

5. _____ *True or False?* Spread footings are typically formed with framing lumber before the excavation for the foundation. (2.3-2)

Lesson 2.4

1. _____ *True or False?* Admixtures might be added to concrete to change its color. (2.4-1)

2. _____ *True or False?* Tensile strength is a material's ability to resist deforming when subjected to forces of compression. (2.4-1)

3. _____ What is the diameter of a No. 3 bar of rebar? (2.4-1)
   A. 1/8″
   B. 3/8″
   C. 1/3″
   D. 3″

4. _____ *True or False?* Two wrenches must be used to properly tighten a carriage bolt. (2.4-2)

5. _____ An anchor bolt would be used to _____. (2.4-2)
   A. secure a cabinet to a wall
   B. mount a microwave oven
   C. fasten a sill plate
   D. clamp a form used for a footing

Lesson 2.5

1. _____ *True or False?* It takes two people to operate an auto level. (2.5-1)

2. _____ The most basic point laser projects _____. (2.5-1)
   A. one point horizontally
   B. one beam vertically
   C. two beams horizontally
   D. two beams vertically

3. _____ Which tool is used to spread and move wet concrete as it is being poured? (2.5-1)
   A. Bull float
   B. Concrete float
   C. Concrete placer
   D. Screed

4. _____ *True or False?* A concrete edge is used to smooth ridges and fill voids left from screeding. (2.5-1)

5. _____ An angle grinder is commonly used to _____. (2.5-1)
   A. cut steel
   B. mix mortar in a wheelbarrow
   C. shape mortar joints
   D. trim the ends of lumber

Lesson 2.6

1. _____ *True or False?* The designated competent person on a jobsite must be either the foreman or owner of the company. (2.6-1)

2. _____ Which of the following is one of the Focus Four hazards? (2.6-2)
   A. Drowning
   B. Electrocution
   C. Poisoning
   D. Suffocation

3. _____ *True or False?* An open manhole cover is an example of a caught-in hazard. (2.6-2)

4. _____ *True or False?* Machines such as excavators and skid steers restrict the operator's visibility dramatically. (2.6-3)

5. _____ When working in trenches, which is the most likely hazard associated with this task? (2.6-3)
   A. Cave-ins
   B. Drownings
   C. Electrocutions
   D. Falls

Name ____

6. _____ A protective system must be used when entering excavations deeper than _____. (2.6-3)
   A. 2′ C. 4′
   B. 3′ D. 5′

Lesson 2.7

1. _____ If there are 8 bolts in a pound, each bolt weighs _____. (2.7-1)
   A. 1/16 lb C. 1/4 lb
   B. 1/8 lb D. 1/2 lb

2. _____ To convert a fraction into an equivalent fraction, you must _____. (2.7-1)
   A. add the same number to the numerator and the denominator
   B. subtract the numerator from the denominator
   C. divide the denominator by the numerator
   D. multiply the numerator and denominator by the same number

3. _____ 1/3 + 3/5 = _____. (2.7-1)
   A. 1/15 C. 14/15
   B. 5/15 D. 1/2

4. _____ 125/116 = _____. (2.7-1)
   A. 1/9 C. 1 1/9
   B. 1 9/116 D. 1 16/25

5. _____ 7/8 – 3/4 = _____. (2.7-1)
   A. 1/8 C. 1/2
   B. 1/4 D. 1

6. _____ *True or False?* In the decimal foot system, 9″ is equivalent to 0.75′. (2.7-2)

7. _____ 16.45′ + 14.67′ = _____. (2.7-2)
   A. 20.02′ C. 30.12′
   B. 21.02′ D. 31.12′

8. _____ Which operation is used to find the difference in elevation from one floor system to the next? (2.7-2)
   A. Addition C. Multiplication
   B. Subtraction D. Division

9. _____ *True or False?* Surveyors use a measuring tape graduated in feet and fractions of a foot. (2.7-2)

10. _____ *True or False?* On an engineer's rod, there are five marks between each tenth of a foot. (2.7-2)

11. _____ When using an engineer's rod, a reading that has six feet, three tenths, and four hundredths is written _____. (2.7-2)
   A. 6.34″ C. 6.30400″
   B. 6.34′ D. 6.30400′

Lesson 2.8

1. _____ Residential plan views are generally drawn to which scale? (2.8-1)
   A. 1/10″ C. 1/4″
   B. 1/8″ D. 1/2″

2. _____ *True or False?* If a certain part of the structure needs to be shown in greater detail in the plans, they are drawn to a larger scale. (2.8-1)

3. _____ *True or False?* All dimensions on construction drawings, even those longer than a foot, are written in inches. (2.8-2)

4. _____ *True or False?* After the excavation is completed, the builder references the foundation plan to establish the size and location of each footing based on the location of the foundation walls. (2.8-3)

Lesson 2.9

1. _____ *True or False?* Reducing building costs is the key consideration for green building. (2.9-1)

2. _____ *True or False?* In the northern hemisphere, a home with its longest axis oriented in an east-west direction will absorb more winter sun on the south side. (2.9-2)

3. _____ An insulated concrete form is made of what material? (2.9-3)
   A. Aerated concrete
   B. Corrugated cardboard
   C. Industrial-grade aluminum
   D. Light polystyrene foam

4. _____ *True or False?* A benefit of using precast walls in green building is that they have a continuous layer of insulation built in. (2.9-3)

Name ______________________ Date ______________ Class ____________

Activity 2-2: Skill Building—Finding Elevations

Learning Outcomes

After completing this lab activity, you will be able to:

- Set up and adjust a builder's level.
- Accurately read the height measurements from a builder's rod.
- Interpret the readings from a builder's rod.

Introduction

One of the first tasks to complete for a construction project is to determine the elevations of site elements and the structure. This is done by an engineer or by the general contractor.

A benchmark is chosen first. A benchmark is a specific spot that will not be disturbed by any construction processes and from where all measurements will be compared by all trades. This is to ensure that all work being done will match up when completed. When a construction worker is checking an elevation, the readings from the level and the rod must be accurate. The calculations required must be done correctly so mistakes do not slow progress.

Instructor Preparation

Instructors must set up a benchmark for the students to work from. Choose a visible place on a wall or column to fasten a block with the top 14″ off the floor. The top of the block is the benchmark height. Have several 4×4 blocks of different lengths ready to set out for the students to take readings from.

Use the following measurements for blocks to start with; others can be added to expand the exercise:

- 4 1/2″
- 9 3/4″
- 12″
- 18 1/2″
- 22 1/4″

Text Reference

Chapter 2, *Excavation and Foundations*

Equipment

The following tools and equipment are needed to perform these activities:

- 4×4 blocks (5) of different lengths
- Builder's level
- Builder's rod
- Carpenter's pencil
- Carpet, triangular piece about 3′ long on each side
- Paper or notebook to record readings
- Safety glasses
- Tripod

Safety Review

Follow all safety procedures specified by your instructor and specified in any appropriate manufacturer's information. The following are some additional safety-related items to keep in mind:

- Wear safety glasses at all times.
- The builder's level is a carefully calibrated instrument and should always be handled carefully.
- Lay the builder's rod down when not in use.

Procedure

1. Set up the builder's level in a location so the benchmark and 4×4 blocks are all visible. If on a concrete floor, use the carpet triangle under the tripod to prevent slipping. Secure the instrument to the tripod. Adjust the legs of the tripod to be at a comfortable height to look through the scope without bending too much and so the top of the tripod is close to level.
2. Once set up, the unit must be carefully leveled for accurate readings. Adjust all four leveling screws so they are in contact with the base of the level. Do not fasten tightly yet. Turn the scope so it lines up with a pair of leveling screws at the base.
3. Adjust the leveling screws using your thumb and forefinger of each hand, turning each at the same time. Always turn the screws in opposite directions. When adjusting, your thumbs are either going toward each other or away from each other. Watch the bubble in the leveling vial on top of the scope and adjust until the bubble is centered in the marks on the vial. Typically, the direction the bubble will move matches the direction of the left thumb.
4. Turn the scope so it lines up with the other two leveling screws perpendicular to the first. Make the adjustments the same way here until the bubble is centered in the vial.
5. Now make sure the leveling screws are set tight to the plate to maintain the level setting. Double check for level.
6. Turn the scope back over the first two screws and check for level again. As adjustments are being made in either direction, the other direction may be affected and may need adjustment. Repeat this step a couple of times to verify the scope is level in all directions. It is now ready to make some readings.
7. Before starting to take readings of the heights of the blocks set out, make sure the crosshairs are visible in the scope. Look through the scope and adjust the eyepiece by turning it until the crosshairs are clearly seen in the view. Now it is ready for measurements.

NOTE

The builder's rod is divided into feet, inches, and eighths of an inch. The top of the line next to a number is the reading for that number. Between each number, the space between marks is 1/8″ and each mark is 1/8″, **Figure 1**. Read the scale just like you would a tape measure and look at the top or bottom of a mark. If the crosshairs are in the middle of any space or line, it can be read using 1/16ths for the measurement.

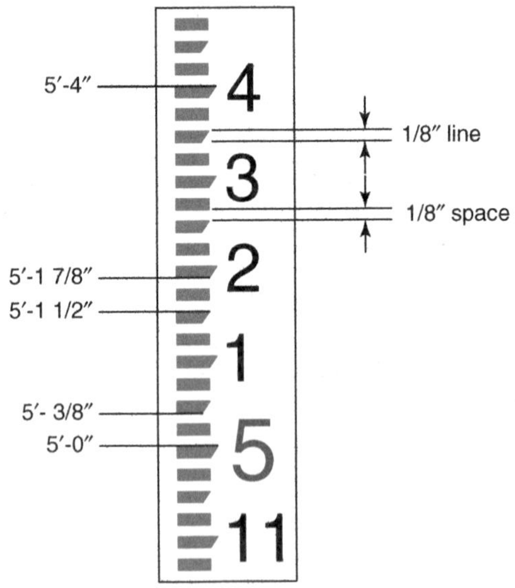

Goodheart-Willcox Publisher

Figure 1

Name ____

8. On a separate sheet of paper, create a four-column table to record the rod reading, elevation difference, and print label for each block. Refer to **Figure 2** for an example.

Sample Block	Rod Reading	Elevation Difference	Print Label
Block #1			

Goodheart-Willcox Publisher

Figure 2

9. When ready to take a reading of the rod, do not try to find the rod while looking through the scope. First, look across the top of the scope and turn it to line up with the rod as close as possible. Then look through the scope to see the rod.
10. When taking a reading at different distances, the focus will need to be adjusted to clearly see the rod. The focusing knob is on either side of the scope near the middle. Rotate this knob to get a clear view of the rod. The crosshairs will remain clear even though the focus of the scope is changed.
11. Take a reading at the benchmark first and record it in the table you sketched on a separate sheet of paper. While using the level, a helper must hold the rod on the measuring location. To be accurate, the helper must hold the rod as plumb as possible while the reading is taken.
12. Ask the helper to hold the rod for a reading at each block. Record each reading in the table in the corresponding space for each block.

NOTE

It may be confusing at first. The builder's rod starts at zero at the bottom of the rod, so a higher reading means a lower elevation, and a lower reading means a higher elevation.

13. When you have finished recording all the readings, remove the level from the tripod. Loosen all four of the leveling screws so they are off the base and then return the level to its proper storage box. Loosening the leveling screws will avoid affecting the calibration while traveling with the level in the box.
14. Next, complete the calculations on the table. Calculate the difference between the benchmark and each reading. This will be the change in elevation in relation to the benchmark. To do this, always subtract the rod reading at each location from the benchmark reading. If the resulting number is negative, the elevation of that location is lower than the benchmark. If the number is positive, the elevation is higher than the benchmark.
15. On a blueprint from the architect, elevations are typically labeled from a benchmark which is designated at 100′ to help make elevation calculations a bit easier. This 100′ label is treated like an elevation above sea level and all other elevations are based on that. If an elevation is 1′ lower than the benchmark, it would be labeled 99′ on the print. Likewise, if an elevation is 2′ higher than the benchmark, it would be labeled 102′. In the last column of the table, calculate what each location would be labeled on the blueprint if the benchmark is 100′. Refer to **Figure 3** for an example.

Sample Block	Rod Reading	Elevation Difference	Print Label
Block #1	5′-9 1/4″	-5 3/4″	99′-6 1/4″

Goodheart-Willcox Publisher

Figure 3

16. Ask your instructor to confirm the calculations are accurate.

Review and Analysis

1. On the builder's rod, what size is the space between lines or each line itself?

____

____

____

2. If there are no crosshairs visible in the scope when looking through it, how would they be made visible?

____

____

____

3. A carpenter was checking the elevations of a yard near a house to plan for an addition. The readings taken as the rod was moved away from the house were as follows: 5′-6″; 6′-2″; 7′-10″; 8′-4″. Does the yard slope downhill or uphill from the house?

____

____

____

Name ______________________ Date ______________ Class __________

Activity 2-3: Skill Building—Drawing a Slab Foundation Plan

Learning Outcomes

After completing this lab activity, you will be able to:

- Identify when to use different types of lines in blueprints.
- Draw a simple foundation plan.
- Draw exterior and interior footing details.

Introduction

To draw a neat and professional set of plans, keep your hands clean of any graphite or lead to avoid smudging, use a sharp pencil for crisp lines, and use a ruler for straight lines. While drawing plans, construction lines and measurement lines will be used to layout the object being drawn. To complete the drawing, object lines and hidden lines will be darkened to the proper line weight, and extension, dimension, and center lines will be added to define the information required to build.

There are different types of foundations available in residential construction, but this activity will focus on slab foundations. Foundation plans are typically drawn to a scale of 1/4″ = 1′-0″ and the footing details will be drawn to a scale that will clearly show the details needed for construction.

Text Reference

Chapter 2, *Excavation and Foundations*

Equipment

The following tools and equipment are needed to perform these activities:

- 45° drafting triangle
- Architect's scale or ruler
- Circle template or small compass
- Eraser
- Erasing shield
- Graph paper, two sheets sized 8.5″ × 11″ with 1/4″ squares
- Masking or painter's tape, 3/4″
- Pencil, standard 2H or HB

Procedure

1. Tape one piece of graph paper to your drawing surface in a portrait orientation.

Foundation Outer Perimeter

The foundation being drawn is 15′ × 10′ with a 12″ perimeter footing and a 12″ center footing. For this activity, as experience is gained, use the 1/2″ = 1′-0″ scale to determine all the slab dimensions while drawing. Actual dimensions or "real size" dimensions will be given in the instructions for locating the extension and dimension lines.

It's important to start drawing your foundation plan using construction lines (light lines) that can be erased or darkened easily, **Figure 1**. After the foundation is drawn completely, then you will go back and darken the lines to their proper line weight. Never use an architect scale as a straightedge to draw lines. The markings along the edge usually have notches which will rub the pencil and create dust from the graphite and will also cause smudges on the paper quickly.

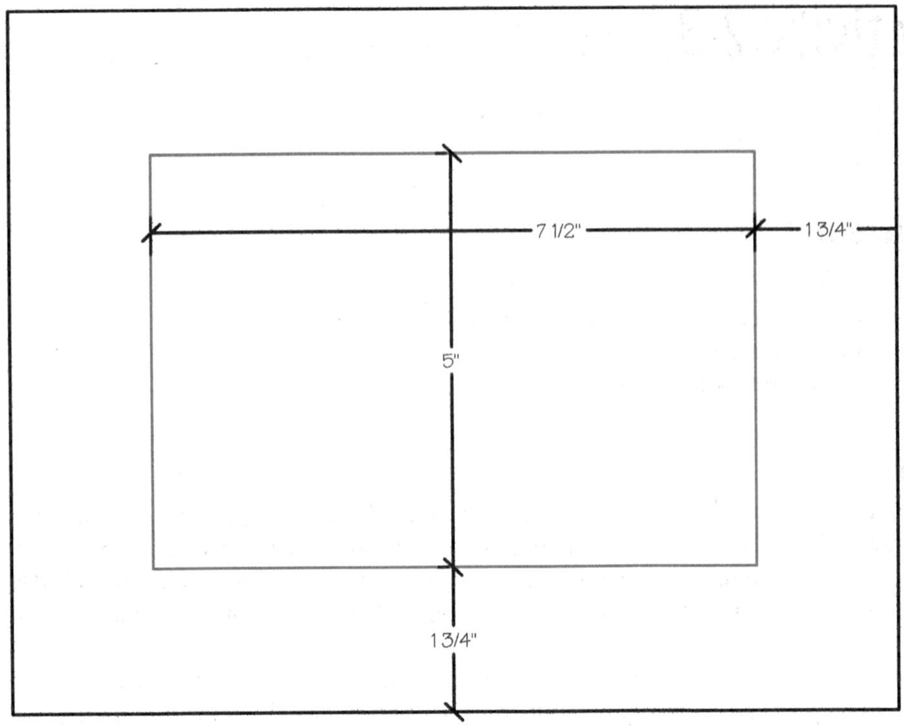

Goodheart-Willcox Publisher

Figure 1

2. For the right side, measure 1 3/4″ from the right side of the graph paper and draw a light vertical line.
3. For the bottom edge, measure 1 3/4″ from the bottom of the paper and draw a light horizontal line.
4. For the left side, using the architect's scale, measure 15′ from the line drawn in Step 2 and draw a light vertical line.
5. For the top edge, using the architect's scale, measure up 10′ from the line drawn in Step 3 and draw a light horizontal line.

Name ______________________________

Interior and Exterior Footings Outlines

This foundation plan will show views of the exterior and interior footings and will need to be broken into two equal parts, **Figure 2**.

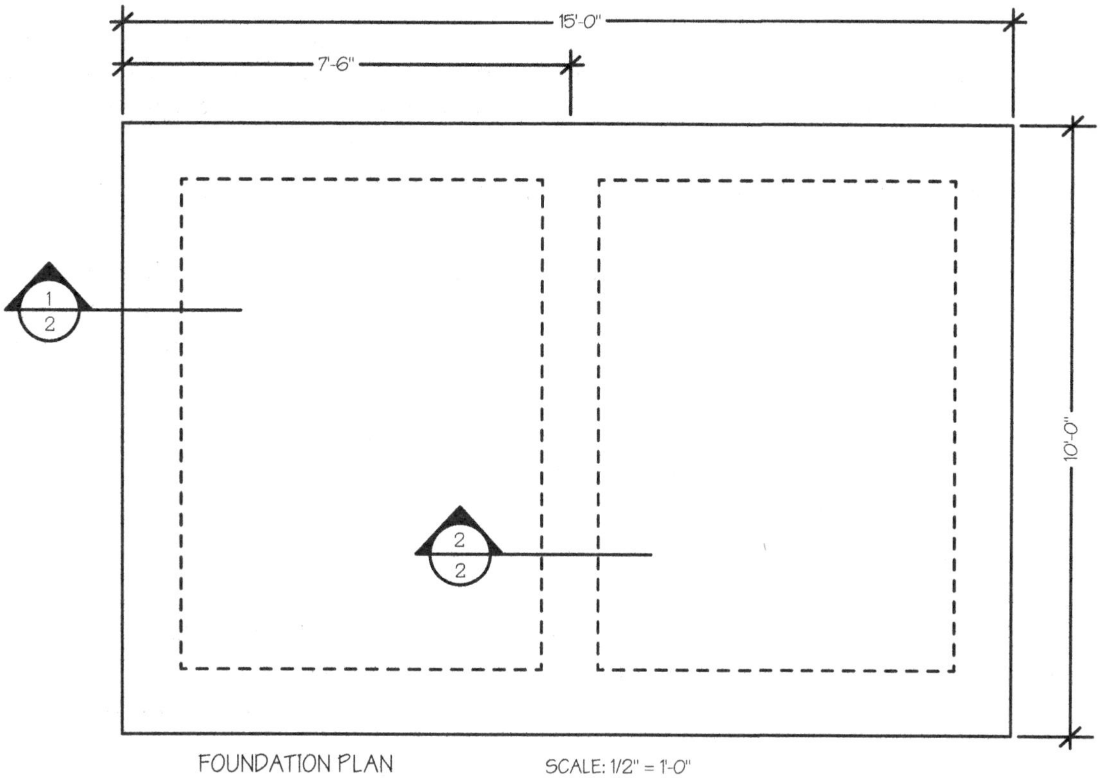

Goodheart-Willcox Publisher

Figure 2

6. To lay out the widths of the exterior and interior footings, find the center of your foundation perimeter and mark it with a light vertical line.
7. From the outside border lines drawn in the *Foundation Outer Perimeter* section, measure in 12″ from all four edges and mark lines to locate the inside of the interior and exterior footings. Draw these lines lightly as they will become hidden (dashed) lines later.
8. From the vertical center line in the middle of your paper, measure 6″ to the left and draw a light line from the top to bottom of your foundation plan. Repeat on the right side; measure 6″ from the center to the right, and then draw a light line from the top to bottom of your foundation plan. These lines locate the bottom of the interior footing and will eventually become hidden (dashed) lines.
9. Darken the construction lines that form the outside border of the foundation into object lines. These should be the darkest lines on the drawing.
10. Redraw all the inside lines as hidden lines with 1/4″ dashes. Draw these lines a little lighter than the object lines but darker than the dimension lines.

Section Cutting Lines

11. To draw section arrows, draw one 2″ long (real size) line through the exterior footing and another 2″ line through the interior footing.
12. To the left of these lines, draw a 1/2″ diameter circle centered vertically.
13. Use the drafting triangle and draw a 45° triangle on the top half of the circle.
14. Darken the area of the triangle that extends outside the circle.

15. Label the section arrows you just created. The top number is the drawing number, the bottom number is the page number.

Dimension and Extension Lines

16. Draw dimension lines at the top of your foundation plan. From the top left corner, measure up 1/2″ (real size) and draw a light line halfway across the top of the foundation to dimension the center footing. Then, measure and draw another 1/2″ light line up from the line you just created all the way across the foundation for the overall length dimension.
17. From the top right corner, measure and draw a vertical line 1/2″ from the right perimeter of the foundation down to the bottom for the overall width dimension.
18. From the two top corners of your drawing in line with the perimeter of the foundation, draw light lines extending 1/8″ past your dimension lines. Start these lines 1/16″ away from your drawing. These will be your extension lines.
19. Do the same for the extension lines on the right.
20. Draw one more extension line from the center of your interior footing to the first dimension line along the top of the foundation. This is a center line, so it will be drawn down into the center footer 2″ (real size). Draw it with a long dash, then a short dash, and then another long dash to meet the dimension line.
21. In the middle of the dimension line or just above the center of the dimension line, add the measurements (dimensions) of the foundation.
22. Darken your dimension lines and extension lines only slightly as these should be the lightest lines on the drawing. Add a tick mark (forward slash) at the beginning and end of each dimension line 1/4″ long.

NOTE

Dimensions locate the width and length of the foundation as well as any features such as the location of interior bearing walls. Any other information is in the section drawings for footings.

Footing Details

To complete the foundation plan, the two detail sections of the footings labeled with the section arrows must be drawn. These detail drawings contain information about the materials being used to construct the foundation. To save space on the drawing, section symbols are used to label some materials instead of a written note. The symbols used in **Figure 3** are for framing lumber (sill plate), concrete (footer and slab), sand and earth (below the concrete slab). The anchor bolts are drawn with dotted hidden lines since they are inside the wall and the concrete and not visible.

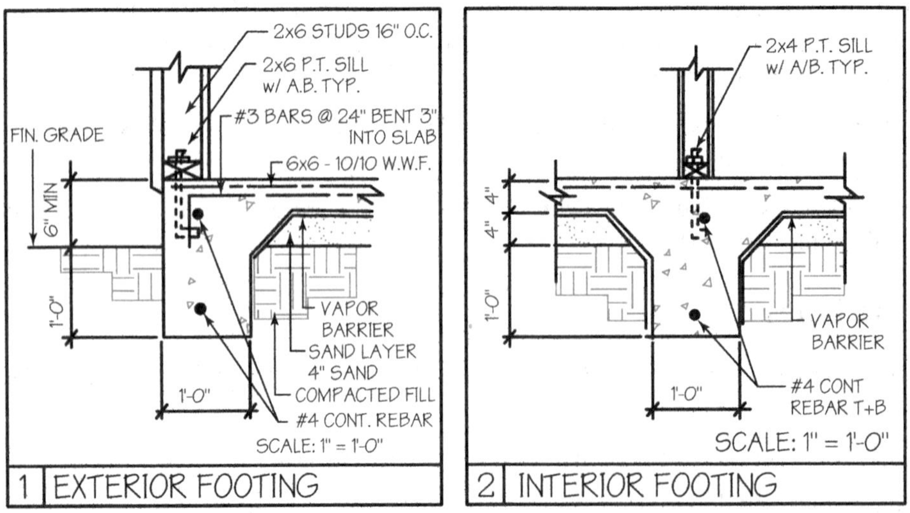

Goodheart-Willcox Publisher

Figure 3

23. On the second piece of graph paper, draw two boxes, filling up the page using dark border lines.
24. Draw each detail of the footer using the 1″ = 1′-0″ scale. Lay out the concrete and footer lines first with light construction lines approximately centered in each box. Match the dimensions and drawings of **Figure 3**.
25. Draw the walls on the concrete slab. The lines only need to be about 1″ (real size) above the concrete and then capped off with the break line. This tells a builder the wall continues up but is not drawn fully in this view. The same is true for the sides of the drawing, a break line shows the same slab construction continues further but is not shown to keep the drawing small.
26. At this point, sketch in all the section symbols showing the materials used. Use **Figure 3** as a guide to draw these. The symbols should be lighter lines.
27. Add in the extension lines and dimension lines defining the sizes of each item in the drawing. Dimensions are always oriented either from the bottom of the page for horizontal dimensions and from the right side of the page for vertical dimensions. A builder should not have to keep spinning a drawing around to read the dimensions.
28. Very lightly and in an organized layout, draw the guidelines for the lettering of the notes. Write in the notes using all capital letters to keep letters uniform and clearly written.
29. Then add leader lines. These are the lines showing where the note applies to the drawing. Each line is drawn with straight segments and a small arrow at the end. Never cross leader lines.
30. Finally, darken the object lines so they stand out from the other lines and symbols. The object being drawn should be the prominent part of a drawing.

Review and Analysis

1. List all the different line types used in foundation plans. Draw an example of each, showing the changes in line weight.

____

____

____

2. What does a break line tell the builder when they are reading the print?

____

____

____

3. Why are the object lines not drawn to the full darkness until the end of the drawing process?

____

____

____

Notes

Name ______________________ Date ______________ Class ____________

Activity 2-4: Skill Building—Laying Concrete Block

Learning Outcomes

After completing this lab activity, you will be able to:

- Lay a concrete block corner lead.

Introduction

The masonry trade is a major part of the construction industry. Masons lay concrete block walls and install brick and stone veneers both inside and outside of buildings. Concrete block walls are built for foundations and fire separation within larger structures. When a mason is building a block foundation or wall, the corners are started first using a mason's level to make them level and plumb. Blocks are then laid between corners using a string line. This activity will give you a chance to lay a concrete block corner lead.

Text Reference

Chapter 2, *Excavation and Foundations*

Equipment

The following tools and equipment are needed to perform these activities:

- 2′ and 4′ levels
- 8″ concrete blocks: 3 corner blocks and 9 stretcher blocks
- Brick hammer
- Carpenter's pencil
- Chalk line
- Jointer
- Mason trowel
- Mortar hoe
- Mortar mixing tub
- Mortar pan
- Safety glasses
- Sand and hydrated lime for mortar
- Shovel

Safety Review

Follow all safety procedures specified by your instructor and specified in any appropriate manufacturer's information. The following are some additional safety-related items to keep in mind:

- Wear safety glasses at all times.
- Wear gloves to protect from rough edges and the chemicals found in mortar.

Procedure

1. Using the 3-4-5 method of squaring a corner, lay out a 90° corner on the floor and snap a chalk line in each direction about 5′ long.
2. Prepare the mortar. Use a mixture of three parts sand and one part lime. Scoop three shovelfuls of sand into the mortar tub. Be sure to keep an equal amount of sand in each shovelful to ensure the correct mix. Then, add one shovelful of lime to the tub. Be sure this scoop is the same size as the ones done with the sand. Mix the dry ingredients together with the mortar hoe before adding water. The lime helps make the mortar stick to the block.
3. Add water to the tub and mix until the mortar has a consistency like toothpaste. It should move around easily but also maintain some shape if made into a pile. If it gets too runny, add a little lime to bring it back to a paste-like consistency.

4. The first block to lay will be one of the corner blocks with a square end. When laying block, be sure to identify which side of the block is the top and which is the bottom. This will help lay to it correctly. The top of a block is smooth and wider, giving a good surface to hold the bed of mortar as each course is laid. The bottom edges have a rough surface and are thinner in width.
5. Set the corner block at the outside of the corner with the outside edge in line with the chalk line, **Figure 1**. This step will establish the width to place the first lines of mortar.

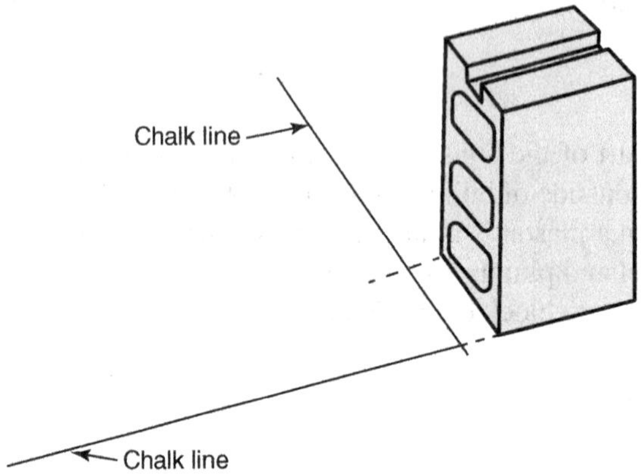

Goodheart-Willcox Publisher

Figure 1

6. Put a few shovelfuls of mortar on a mortar pan or a piece of plywood near the work area. Using the trowel, scoop the mortar into a long pile. Take the trowel and scoop up some mortar so there is line of mortar along the length of one edge, **Figure 2**.

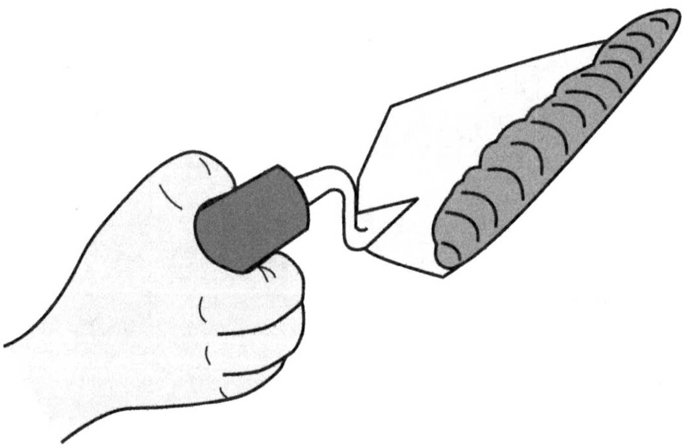

Goodheart-Willcox Publisher

Figure 2

7. Place the mortar on the floor next to the chalk line at the outside edge of the block creating a long pile about 1″ high. Be sure the chalk line is still visible. Do this a few times until there is a line of mortar a bit longer than the length of one block.
8. Repeat this process in line with the inside edge of the corner block to be laid. Make this line of mortar the same length as the first. There should now be two lines of mortar that line up with the width of the block.

Name ______________________________

9. Place a small scoop of mortar across the end of the lines at the corner for a full bed joint. See the following figure showing the mortar ready for the first block.

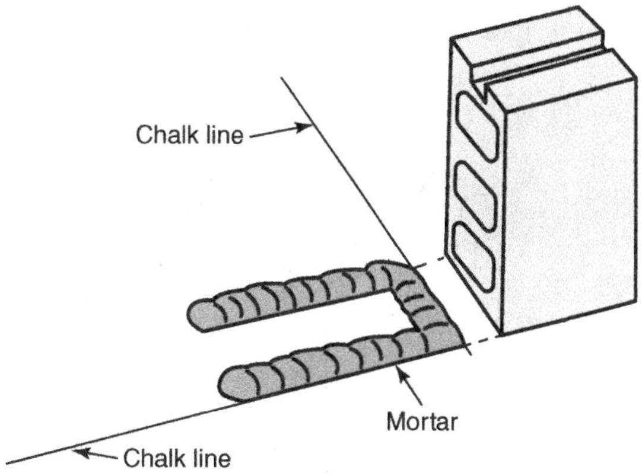

Goodheart-Willcox Publisher

Figure 3

10. Pick up the block from the top and place it slowly into the lines of mortar. Look at both outside edges of the block, making sure they are in line with the chalk lines on the floor. Let go of the block when it has settled enough in the mortar to stay in place.
11. Check the block for level by placing the 2′ level on top. Check along the length of the block and then along the width of the block. Observe where it is high and tap the block with the brick hammer where it needs to be lowered. Check the level in both directions again and use the hammer again if needed to finish making it level. Then, check the block at the corner for plumb. Refer to **Figure 4** for how to check for plumb.

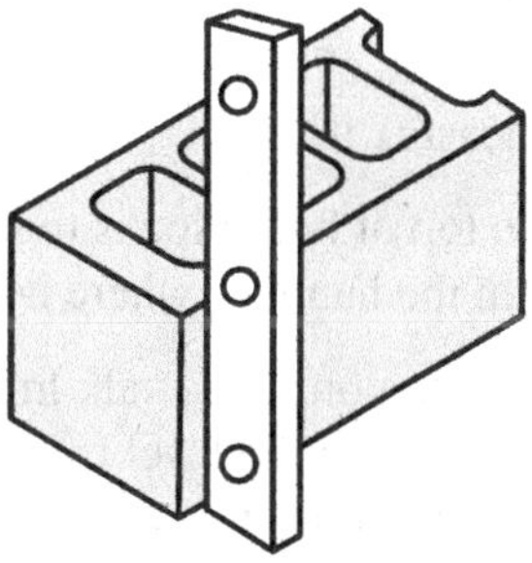

Goodheart-Willcox Publisher

Figure 4

12. Clean the mortar off the floor and along the outside of the block; then, check if the block is still lined up with the chalk line in both directions. If needed, tap the block one way or the other with the hammer to get the end and sides of the block directly over the lines. Clean the mortar along the inside of the block as well and place the excess mortar back on the pan to be used again.
13. Check the height of the block. Measure from the floor to the top of the block. Concrete block is made 3/8″ smaller than the nominal size of the block to allow for the mortar joint between each block. The top of the block should be 8″ high with the proper mortar bed under it. If it is too high, tap around the top of the block to bring it down and then double check the level in both directions.
14. Place more mortar down along the chalk line, creating two rows spaced the same width as a block. Go far enough for two more blocks.
15. Take two stretcher blocks and set them next to the rows of mortar on their ends with the top of the block facing away from the corner.
16. Pick up a slice of mortar on the edge of the trowel. To get the mortar to stick to the trowel for the next procedure, give the trowel a little snap of the wrist downward.

17. Turn the trowel over and starting with the edge of the trowel with no mortar on it, slide it down the top corner of the block, depositing the mortar on the ear of the block. Pick up another slice of mortar on the opposite edge of the trowel and slide it down the corner of the block to deposit it on the other ear of the block. Run the edge of the trowel along the inside of the ridge of mortar to push it into the block to help it stick. The block should resemble **Figure 5**.

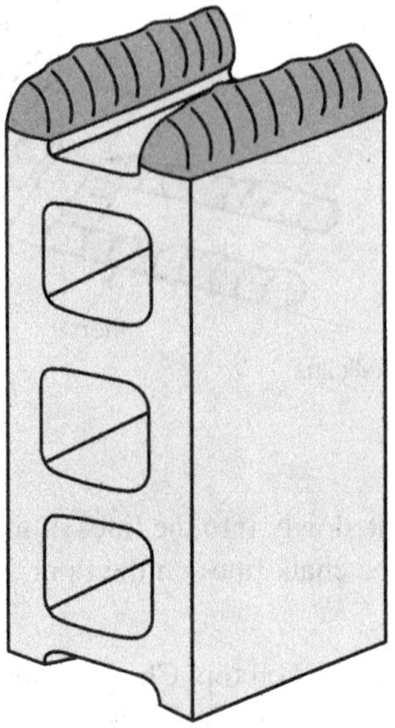

Goodheart-Willcox Publisher

Figure 5

18. Next, pick up the block with the mortared end toward the block already laid. Set the block down into the mortar bed in line with the chalk line and against the end of the first block, making sure to squeeze the mortar a bit. Again, let it settle into the mortar before releasing it.
19. Now take the 4′ level and check the level on top of both blocks together. Check the level across the second block making sure the two blocks line up. Tap with the hammer where needed to make both blocks level and plumb.
20. Repeat the same process and lay the third block along the chalk line. Using the 4′ level, hold it along the sides of all three blocks ensuring they are in line. Then, slide the level farther down the row and toward the floor to make sure they are in line with the chalk line as shown in **Figure 6**. Adjust if needed.

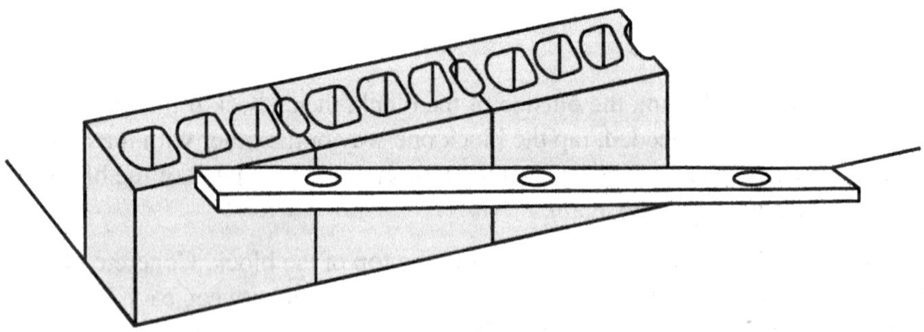

Goodheart-Willcox Publisher

Figure 6

21. Lay two more blocks along the chalk line in the other direction to complete the corner. Check all for level and plumb in both directions, and make sure the blocks are in line with the chalk line.
22. Using the technique practiced in Steps 16 and 17, place a ridge of mortar along the top of both edges of the blocks already laid on the floor. In the second course, place the corner block so it is perpendicular to the first block, overlapping the first head joint. Set the block in the mortar bed, keeping it in line with the block below. Once settled in the mortar, check it for level in both directions and check the corner for plumb on both sides of the corner. Adjust as needed with the hammer or the handle of the trowel.

Name ____

23. Once it is level and plumb, use the trowel to scoop any excess mortar off the sides of the block at the joints and return it to the mortar pan for reuse.

24. Complete the second course with one block in one direction and two blocks in the other. Check for level in both directions and check for plumb on the outside of the corner. Use the 4′ level to check alignment with the chalk line as was done in Step 20. Move the level up to the second course for this step.

25. Repeat these steps to complete the corner lead with a third course to match **Figure 7**. Check the height of the third course. The top of the block with the correct size mortar joints should now be 24″ above the floor.

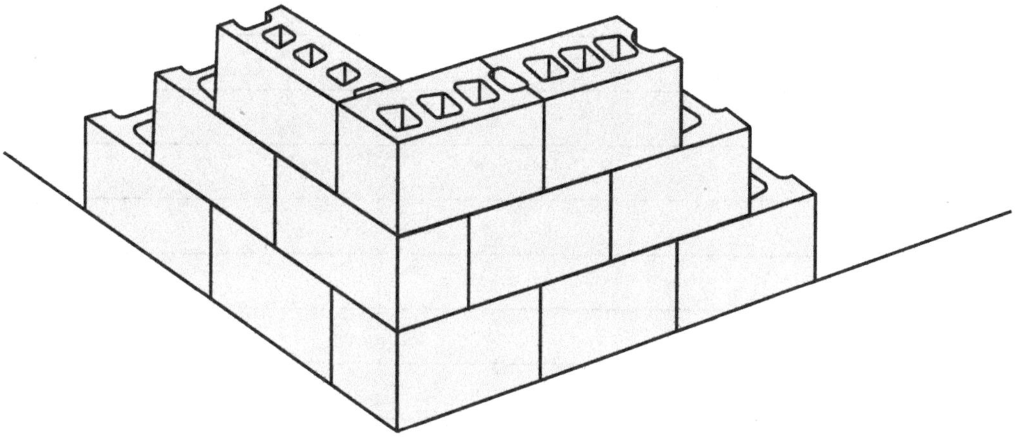

Goodheart-Willcox Publisher

Figure 7

26. Before the mortar sets too hard, use the jointer to smooth the mortar by rubbing the curved end along each joint. This helps compress the mortar in the joint and gives it a finished appearance. Start with the joint along the floor and the inside corner. If a spot needs a little more mortar to fill in smoothly, use the trowel to bring some mortar to the joint and use the jointer to slide it off the trowel and into the joint.

Review and Analysis

1. Describe the purpose of the wrist snap when loading the trowel. What do you think would happen to the mortar if it were not snapped firmly enough?

____

____

____

2. What characteristic(s) of the mortar would change if more lime were added to the standard mix?

____

____

____

3. Which should take precedence as more courses are added to the corner: checking each corner block is level or that the corner being built is plumb? Explain why.

____

____

____

Notes

Activity 2-5: Project—Forming and Pouring Concrete Stepping Stones

Learning Outcomes

After completing this lab activity, you will be able to:

- Build a concrete form using the proper materials.
- Explain how the materials and measurements needed to make concrete impact the mixture.
- Demonstrate how to mix and pour concrete.

Introduction

In this activity, you will be creating a concrete form from lumber, and then mixing and pouring concrete to create a square stepstone that measures 15″ × 15″ × 2″. This type of concrete form is also used to create column footings but would need to be much larger and deeper in order to support the weight of the structure. The size of a footing is designed by an architect or engineer and depends on the type of soil they're building on as well as the weight of the structure being supported. Concrete is created using a mix of materials: aggregate, construction sand, cement, and water. After the concrete is poured, you will need to level and smooth the surface and finish the edges.

Text Reference

Chapter 2, *Excavation and Foundations*

Equipment

The following tools and equipment are needed to perform these activities:

- 6′ of 1×4 lumber
- Carpenter's pencil
- Circular saw or power miter saw
- Clean rag or 2″ paint brush
- Clean water
- Common nails (12), size 8d
- Concrete edging tools, 1/2″ radius edger and hand groover
- Concrete float
- Concrete hoe
- Construction sand
- Drill with #2 screwdriver tip to match screws
- Drywall screws (6), size 1 1/4″
- Framing hammer
- Linseed oil or release agent
- Pea gravel
- Plywood measuring 3/4″ × 15″ × 15″
- Safety glasses
- Screed measuring 2″ × 4″ × 18″
- Small concrete mixing tub
- Small shovel
- Soft broom or medium brush
- Spacers (2) measuring 3/4″ × 3″ × 14″
- Speed square
- Tape measure

Safety Review

Follow all safety procedures specified by your instructor and specified in any appropriate manufacturer's information. The following are some additional safety-related items to keep in mind:

- Wear safety glasses at all times.
- Wear gloves when working with materials that may damage your hands.
- Make sure your tools are clean before you start.

Procedure

1. Crosscut your 1×4 into 4 pieces 18″ long using the circular saw or power miter saw.

> **NOTE**
> When larger forms are needed for more concrete, you would likely need lumber that matches the depth of the building component being created with stakes to brace the frame so the lumber does not bow from the weight of the concrete.

2. Nail the 1×4s together to create a square that measures 15″ × 15″ inside the form, **Figure 1**.

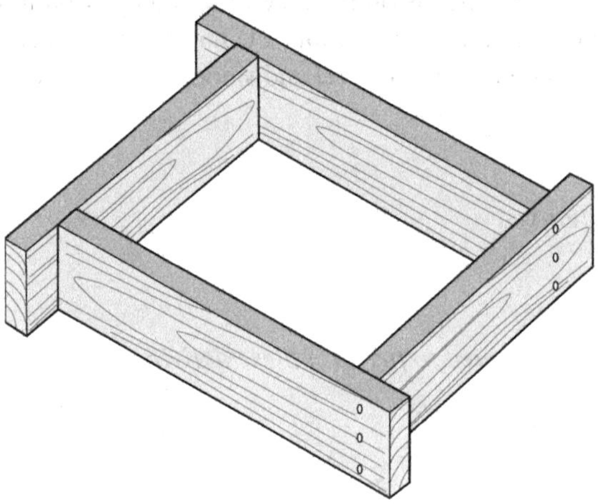

Goodheart-Willcox Publisher

Figure 1

3. Cut the plywood with a circular saw to be 15″ wide. Trim one end square and crosscut to 15″ in length.
4. Screw the spacers to the bottom of the plywood with 1 1/4″ screws. Lay the spacers flat on the floor under each end of the plywood. Line up the edges of the spacer to be flush with the edges of the plywood before fastening. Drill the proper shank hole in the plywood for the screw to have a tight fit.
5. Place the plywood bottom inside the form as a base, **Figure 2**. This should create a step stone that will be 2″ thick.

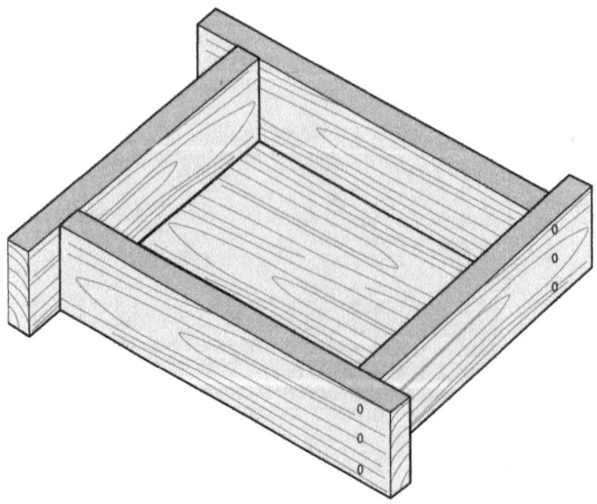

Goodheart-Willcox Publisher

Figure 2

Name ________________________________

6. Apply the linseed or release agent with the brush to the inside of the form. Make sure to coat the sides and the bottom of the form evenly.
7. Gather materials needed for mixing concrete: aggregate (washed crushed rock, pea gravel size), construction sand, cement, and water.
8. In a concrete mixing tub, mix concrete using a seven-part mix consisting of four parts aggregate, two parts sand, and one part cement. Mix the dry ingredients with the concrete hoe thoroughly before adding the water. Add one-half measurement of water and mix thoroughly. Measuring out equal parts of aggregate, sand, and cement can be done with a shovel, bucket, or a metal/plastic can that is similar to a measuring cup.

NOTE

If more water is needed, add a little at a time. Too much water will create weak concrete. The concrete mixture should be stiff, not soupy.

9. Shovel the mixed concrete into the form. A little extra is better than a little less.
10. Remove air pockets and bubbles in the mixture by pushing or poking your shovel into the concrete. Tap around the outside of the form to remove air bubbles against the form boards. Next, use the 2″ × 4″ × 18″ strike board to screed the concrete, **Figure 3**. To screed the form, place the screed board at the top of the form. Move the board in a side-to-side motion while slowly pulling the screed across the top of the form to level the concrete and remove excess mixture. Repeating the process will create a smooth, level surface.

Goodheart-Willcox Publisher

Figure 3

11. Let the concrete set for 15 minutes.
12. Use the concrete float to smooth the surface. Place the float flat on the surface of the concrete and slide it around the top of the form, leveling the surface. Tilt the trowel just a bit at first to make sure it does not dig into the surface, but hold the trowel flat as the surface becomes more even. The aggregate will be pushed down and water will be forced to the top of the concrete, helping to create the smooth surface.
13. Radius the edges using the concrete edger. Push the rounded curve of the edger into the concrete at the edge of the form board and slide it along the edge, pushing the aggregate away from the form board. Keep the edger against the form board and riding on the surface of the concrete. Go back and forth along the form until all the spaces between the aggregate are filled with concrete. Repeat the process for each side of the form.

14. Divide the step stone into quarters using the concrete groover. Clean the screed and use it as a straightedge for the groover. Place the screed board across the slab in position so the center of the groover lines up with the quartered layout marks. While holding the groover against the board, push it into the concrete. This pushes the aggregate out of the way to form the groove. Pick up the edger and move it along the board in steps, pushing it in at each step. Once across the slab, slide it back and forth across the surface to smooth the groove until all the voids around the aggregate are filled. Place the board across the slab in the other direction and repeat.
15. Let the concrete set for another 15 minutes and then use the float, edger, and groover to touch up any areas that are uneven, making sure the aggregate has been pushed down for a smooth surface.
16. Brush the surface lightly with a soft brush or broom to give the surface a non-slip texture.
17. Wash, clean, and dry all the tools you used to mix and handle the concrete.
18. Let the concrete cure for 24 hours before removing the form boards and then clean the form boards.

Review and Analysis

1. What are the ingredients and measurements required to make concrete?

____

____

____

2. What happens when too much water is added to concrete?

____

____

____

3. What are other ways that concrete can be mixed?

____

____

____

Name ______________________ Date ____________ Class ____________

CHAPTER 3 Framing Systems

Activity 3-1: Chapter 3 Review

Answer the following questions after studying the chapter.

Lesson 3.1

1. _____ *True or False?* In a sole proprietorship, the business and the owner are considered the same legal entity. (3.1-1)

2. _____ A business structure in which two or more people own a business together and each contributes to the business through financial investment, skills, and/or labor is called a _____. (3.1-1)
   A. corporation
   B. partnership
   C. limited liability company
   D. sole proprietorship

3. _____ *True or False?* A sole proprietorship offers a higher level of protection for a business owner than a limited liability company. (3.1-1)

4. _____ Which business structure offers the best protection to the business owner? (3.1-1)
   A. Corporation
   B. Limited liability company
   C. Sole proprietorship
   D. Partnership

5. _____ Which statement about the corporation business structure is true? (3.1-1)
   A. All owners of the corporation are actively involved in managing it.
   B. The cost to form a corporation is lower than that of other business structures.
   C. The owners of the corporation are individually taxed for all of its profits.
   D. Ownership of the corporation is allocated through shares.

Lesson 3.2

1. _____ Building codes exist to _____. (3.2-1)
   A. meet federal regulations regarding building construction
   B. identify ways to make a building more energy efficient
   C. help ensure the health and safety of building occupants
   D. reduce the cost of construction for the builder and homeowner

2. _____ A set of building codes that is generally accepted across the industry and can apply to any geographic location is a _____ code. (3.2-1)
   A. general
   B. model
   C. federal
   D. uniform

3. _____ When planning to construct a residence in a city, a formal application for a building permit is typically submitted to the _____. (3.2-2)
   A. city clerk
   B. county clerk
   C. city court clerk
   D. county court clerk

4. _____ To ensures that the depth of excavation meets building requirements, a _____ inspection must be conducted. (3.2-3)
   A. drywall
   B. footing
   C. foundation
   D. site

5. _____ *True or False?* HVAC inspections initially occur before the rough-in stage of construction. (3.2-3)

Lesson 3.3

1. _____ The process of assembling the floor, walls, and roof structures of a building is called _____. (3.3-1)
   A. framing
   B. finishing
   C. rigging
   D. joisting

2. _____ Studs are continuous from the sill to the rafter plate with _____ framing. (3.3-1)
   A. post-and beam
   B. balloon
   C. platform
   D. western

3. _____ The distance between the sills of opposing walls or other structural supports is called the _____. (3.3-2)
   A. plate
   B. seal
   C. truss
   D. span

4. _____ *True or False?* Oriented strand board is commonly used for the subfloor. (3.3-2)

5. _____ A spacing of _____ OC is common for wall studs. (3.3-2)
   A. 14″
   B. 16″
   C. 18″
   D. 20″

6. _____ *True or False?* Rafters serve as the supports for the roof load. (3.3-2)

Lesson 3.4

1. _____ *True or False?* Engineered materials use about 40% of a log. (3.4-1)

2. _____ Which statement about laminated-veneer lumber products is true? (3.4-1)
   A. They are impervious to staining.
   B. They require special tools to cut.
   C. They are dimensionally stable.
   D. They can be used for finishing, but not for framing.

3. _____ *True or False?* Conventional plywood and laminated-veneer lumber have their grain running in the same direction. (3.4-1)

4. _____ Open-web floor trusses are fabricated in factories from solid _____ lumber. (3.4-1)
   A. 2×4
   B. 2×6
   C. 4×4
   D. 4×6

5. _____ Which statement about hangers is true? (3.4-2)
   A. They are designed to be fastened using strong construction-grade adhesives.
   B. They connect framing members together at the center of each member.
   C. They are solid and require holes to be drilled for fasteners before installing.
   D. They are used where floor or ceiling joists intersect another framing member.

6. _____ *True or False?* Cements used in construction are rubber-based. (3.4-2)

7. _____ What are the primary fasteners used when framing a home? (3.4-2)
   A. Screws
   B. Nails
   C. Glues
   D. Adhesives

8. _____ *True or False?* A framing nail has a smaller head than a trim nail. (3.4-3)

Lesson 3.5

1. _____ What is the most often used fastening tool for framing? (3.5-1)
   A. Ball peen hammer
   B. Claw hammer
   C. Wrecking bar
   D. Flat bar

Name ______________________________

2. _____ *True or False?* The ripping claw hammer is the tool best suited for pulling nails. (3.5-1)

3. _____ *True or False?* A pry bar is designed for dismantling assemblies. (3.5-1)

4. _____ Wrecking bars vary in length from _____. (3.5-1)
   A. 2″ to 8″
   B. 7″ to 15″
   C. 10″ to 24″
   D. 12″ to 36″

5. _____ Which statement about a framing square is true? (3.5-1)
   A. The outside edge of its face is divided into 1/8″ graduations.
   B. Its tongue carries the Essex Board Measure Table.
   C. Carpenters commonly use it to find the length and angles of rafters.
   D. Its body is 1 1/2″ wide and 16″ long.

6. _____ *True or False?* Staples generally provide greater holding power than nails. (3.5-1)

7. _____ Which type of saw uses a back-and-forth, linear cutting motion? (3.5-1)
   A. Reciprocating saw
   B. Chop saw
   C. Table saw
   D. Power miter saw

Lesson 3.6

1. _____ What is the most effective control for managing hazards? (3.6-1)
   A. Substitution
   B. Personal protective equipment
   C. Elimination
   D. Administrative controls

2. _____ *True or False?* One important strategy for avoiding struck-by hazards is to always keep an eye on what is going on above you. (3.6-2)

3. _____ OSHA standards require that the footwear worn on construction jobs _____. (3.6-3)
   A. use Velcro straps instead of laces
   B. be brightly colored for high visibility
   C. feature extra-deep treads for superior traction
   D. have reinforced toes that can withstand heavy loads

4. _____ *True or False?* A particulate mask is considered a respirator under OSHA standards. (3.6-3)

Lesson 3.7

1. _____ When reduced to the lowest terms, 24/60 = _____. (3.7-1)
   A. 2/5
   B. 3/8
   C. 4/10
   D. 6/15

2. _____ 4/10 ÷ 3/5 = _____. (3.7-1)
   A. 6/25
   B. 2/3
   C. 1 1/2
   D. 4 1/6

3. _____ *True or False?* When operations will include adding fractions, you must convert the fractions to their least common denominators. (3.7-1)

4. _____ *True or False?* Finding the center of a dimension involves splitting the dimension into two equal parts. (3.7-2)

5. _____ What is the center point of a wall plate that measures 17′-6 3/5″? (3.7-2)
   A. 8′-3 3/5″
   B. 8′-3 3/10″
   C. 8′-8 3/10″
   D. 9′-3 2/3″

6. _____ *True or False?* To double a fraction, you simply cut the numerator in half. (3.7-2)

Lesson 3.8

1. _____ *True or False?* The overall building dimension is typically closest chain dimension to the center of the drawing. (3.8-1)

2. _____ *True or False?* When a dimension line intersects an extension line without an endpoint, it indicates that the dimension continues beyond this extension line. (3.8-1)

3. _____ Center lines are drawn on a plan using a _____. (3.8-1)
   A. thick line of short dashes
   B. thin line of short dashes
   C. thick line of alternating long and short dashes
   D. thin line of alternating long and short dashes

4. _____ *True or False?* The floor plan includes references to additional information such as section and detail drawings. (3.8-2)

5. _____ *True or False?* Dimensions are used on framing plans. (3.8-2)

Lesson 3.9

1. _____ Advanced framing uses _____. (3.9-1)
   A. 2×4 studs spaced 16″ OC
   B. 2×4 studs spaced 24″ OC
   C. 2×6 studs spaced 16″ OC
   D. 2×6 studs spaced 24″ OC

2. _____ *True or False?* When using advanced framing, the studs must be placed directly over the floor joists at each level. (3.9-1)

3. _____ *True or False?* In a building constructed with advanced framing, headers are required for non-loadbearing walls. (3.9-1)

4. _____ Which statement about structural insulated panels is true? (3.9-2)
   A. They are less expensive than the materials used in conventionally-framed walls.
   B. They are perforated to allow sufficient air circulation once installed.
   C. They produce essentially no waste during the framing phase of construction.
   D. They are only appropriate for use in vertical applications, such as walls.

Name ______________________ Date ______________ Class ____________

Activity 3-2: Skill Building—Sketching a Bedroom and Bathroom to Scale

Learning Outcomes

After completing this lab activity, you will be able to:

- Demonstrate the use of line types used in floor plans.
- Identify basic plumbing and electrical symbols used in floor plans.

Introduction

In this activity, you will sketch a scale drawing of your bedroom and bathroom. You will locate doors, windows, and electrical and plumbing features. This activity will help you become more familiar with architectural floor plans.

Line types typically used in floor plans include:

- Object lines—Solid lines used to show the outlines of walls, doors and windows.
- Hidden lines—Dashed lines showing something that is not visible from the viewing position. They will be used in this drawing to show wiring connecting light switches to lights or split duplex outlets.
- Extension lines—Lines that extend from corners or centers of an object to show the beginning and ending of a dimension.
- Center lines—Lines that are made up of two longer lines with a short dash in the center that are used to mark the center of any object like windows and doors.
- Dimension lines—Lines on either side of a dimension that show the extent of the specific measurement being shown. Their ending point is marked with a tick mark or arrow.

Text Reference

Chapter 3, *Framing Systems*

Equipment

The following tools and equipment are needed to perform these activities:

- Eraser
- Graph paper with 1/4″ grid
- Measuring tape
- Pencil
- Ruler or drafting triangle

Procedure

Measure your bedroom and sketch it on graph paper at a scale of 1/2″ = 1′-0″ to create a bedroom floor plan. With 1/4″ graph paper, each square will be equal to 6″. After this, add architectural symbols.

Exterior walls are typically framed with 2×6 lumber and interior walls framed with 2×4 lumber. Walls behind the toilet usually contain larger plumbing pipes and are referred to as plumbing walls. They are usually framed with 2×6 lumber to provide enough room for these pipes to go through the wall. After creating your bedroom floor plan, you will create a bathroom floor plan. Refer to the sample drawings provided. See **Figure 1**.

For this activity, draw all your walls with a thickness of 6″, at scale.

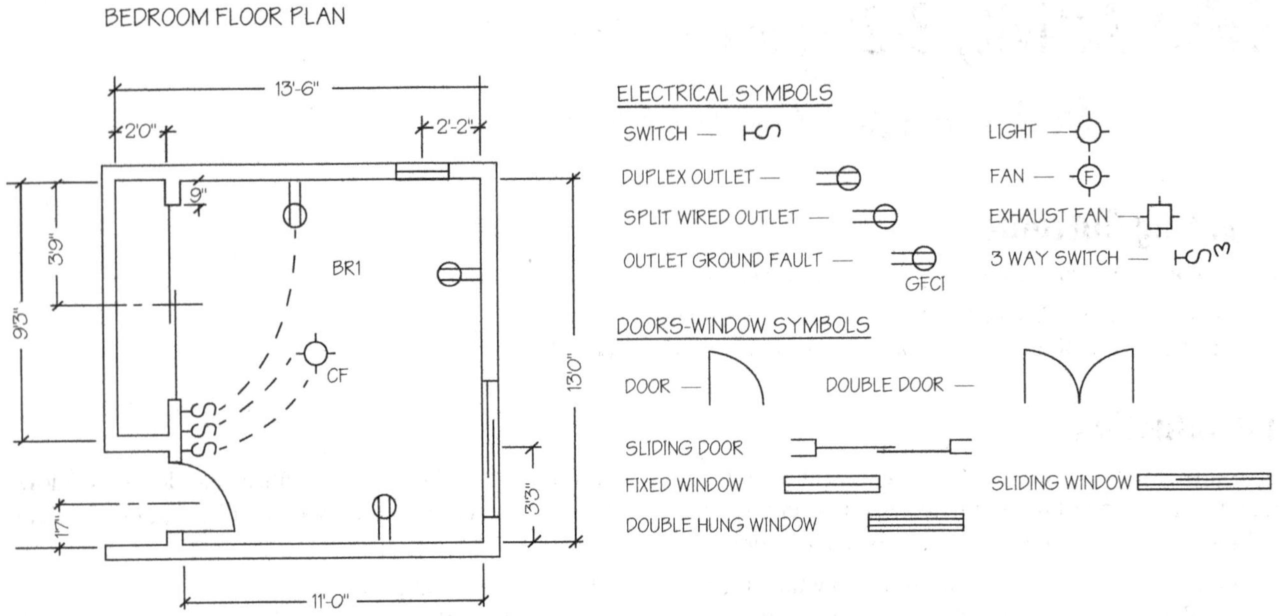

Goodheart-Willcox Publisher

Figure 1

Bedroom

1. Measure your bedroom and sketch the perimeter of the room on graph paper. Use the ruler or drafting triangle as a guide to keep the lines straight and neat.
2. Each wall in your sketch will be 6″ thick. Measure to the outside of each line and duplicate these lines to create your bedroom walls.
3. Measure the width of your doorway. Find the center of your doorway and measure to the closest corner.
4. Lay out the distance measured from the corner to the center of the door on the corresponding wall on the drawing. Divide the width of the door in two and measure to each side of the center mark to locate the door. Draw the proper door symbol on the floor plan and dimension it from the corner of your room to the center of the doorway. Use a center line in the door opening to locate the end of the dimension line.
5. Measure and draw any closet that opens into the bedroom. Dimension the door from the nearest corner to the center of the door as before. Add dimension lines inside the closet to show the size.

NOTE

Place dimensions 1/2″ (real size) away from your wall and other dimension lines.

6. Measure the width of any window(s) and locate their centers. From the center measure to the closest corner.
7. Lay out that length from a corner to locate the center of the window. Lay out half the width of the window on either side of this mark. Draw the window(s) on the floor plan using the proper window symbol and dimension from the corner to the center of each window. Be sure to use the center line at the end of the dimension line in the window opening.
8. Locate all electrical components in your bedroom, such as duplex outlets, light switches, ceiling fans, and lights. Measure how far away they are from each other and one corner of the bedroom.
9. Draw the electrical symbols on your floor plan matching the measurements you collected in the previous step.
10. Draw hidden lines showing which light switches control each of the lights or ceiling fans. Use a dashed hidden line drawn from the switch to the symbol for the device being controlled. Sometimes a wall switch will control half of a duplex wall outlet to turn a light on. This is called a split wired duplex outlet. Draw the hidden line to show that connection from the switch to the outlet.
11. Add any smoke detectors, using a small circle with SD written next to the circle.

Name ____________________

Bathroom

1. Measure your bathroom and sketch on graph paper. Include rooms that have sinks, shower, bathtub, toilet, and cabinets. See **Figure 2**.

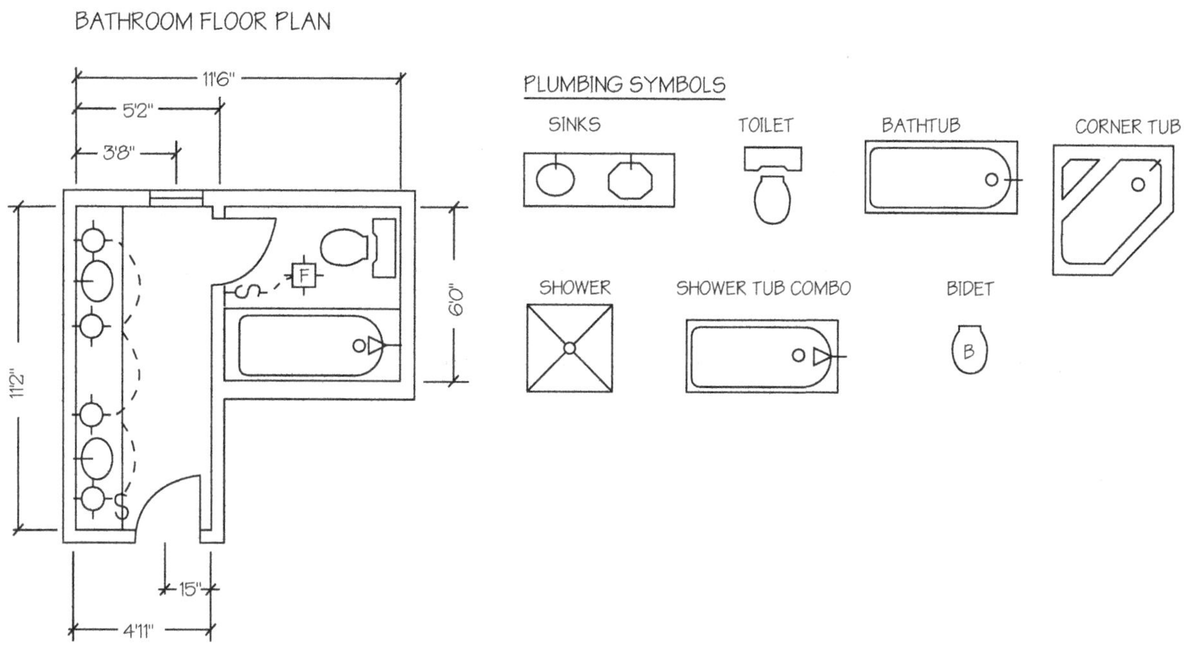

Goodheart-Willcox Publisher

Figure 2

2. Measure the width of your doorway. Find the center of the doorway and measure to the closest corner.
3. Draw the door symbol on the floor plan and dimension it from the corner of the bathroom to the center of the doorway.
4. Measure the width of any window(s) and locate their centers. From the center measure to the closest corner.
5. Draw the window(s) on the floor plan and dimension it from the corner to the center of each window.
6. Locate all electrical components in the bathroom, such as duplex outlets, light switches, and lights. Measure how far away they are from each other and one corner of the bathroom.
7. Draw the electrical symbols on the floor plan matching the measurements you made in the previous step.
8. Draw hidden lines connecting light switches to lights and split wired duplex outlets.
9. Draw in plumbing symbols, such as a bathtub, shower, sinks, toilet, or bidet.
10. Add any smoke detectors using a small circle with SD written next to the circle.

Review and Analysis

1. How far apart are the duplex outlets placed on your bedroom walls?

2. How many square feet is your bedroom?

3. When looking at floor plans, what are object lines and hidden lines used for?

Name ______________________ Date ______________ Class ____________

Activity 3-3: Skill Building—Laying Out Framing

Learning Outcomes

After completing this lab activity, you will be able to:

- Lay out framing members in a floor and wall system.
- Identify when adjustments must be made for structural elements.

Introduction

Laying out the framing members of a house is one of the required skills of a builder. They must read a print and interpret the information to get the framing prepared for the rest of the structure. Not only do they have to understand the structure, but they also much understand other parts of the residence, such as the plumbing and electrical systems. These utility systems must be examined and prepared for as well.

Structurally, framing members should be lined up to create a continuous load path from the roof to foundation. Framing may need to be altered to prepare for plumbing piping and fixtures so framing members are not cut out during installation. When this is done right from the start of the framing process, it will help all other stages of construction.

Instructor's Note: Using a 2″ × 4″ × 12′ lumber for this exercise may be done in several different ways to help with material costs. Mark the lumber lightly so the lines can be erased or sanded off and the lumber can be reused. If cash register tape can be obtained, roll that out and tape that to the 2×4 while doing the layout on the paper, or make marks for the layouts along only one edge of the board and then other layouts may be done on other edges. Understanding the layout process is what's important, not necessarily the marking.

Text Reference

Chapter 3, *Framing Systems*

Equipment

The following tools and equipment are needed to perform these activities:

- 2″ × 4″ × 12′ framing lumber
- Carpenter's pencil
- Combination or speed square
- Safety glasses
- Tape measure

Safety Review

Follow all safety procedures specified by your instructor and specified in any appropriate manufacturer's information. The following is an additional safety-related item to keep in mind:

- Wear safety glasses at all times.

Procedure

1. Use the prints in **Figure 1** and **Figure 2** to collect information about the floor and wall systems being shown. In each print, the labels denote where the layouts will begin. Based on the arrow showing the floor joist direction, Wall A and Wall B of this house are loadbearing. That is where the ends of the joists will be supported and eventually where the rafters or trusses will also be supported.

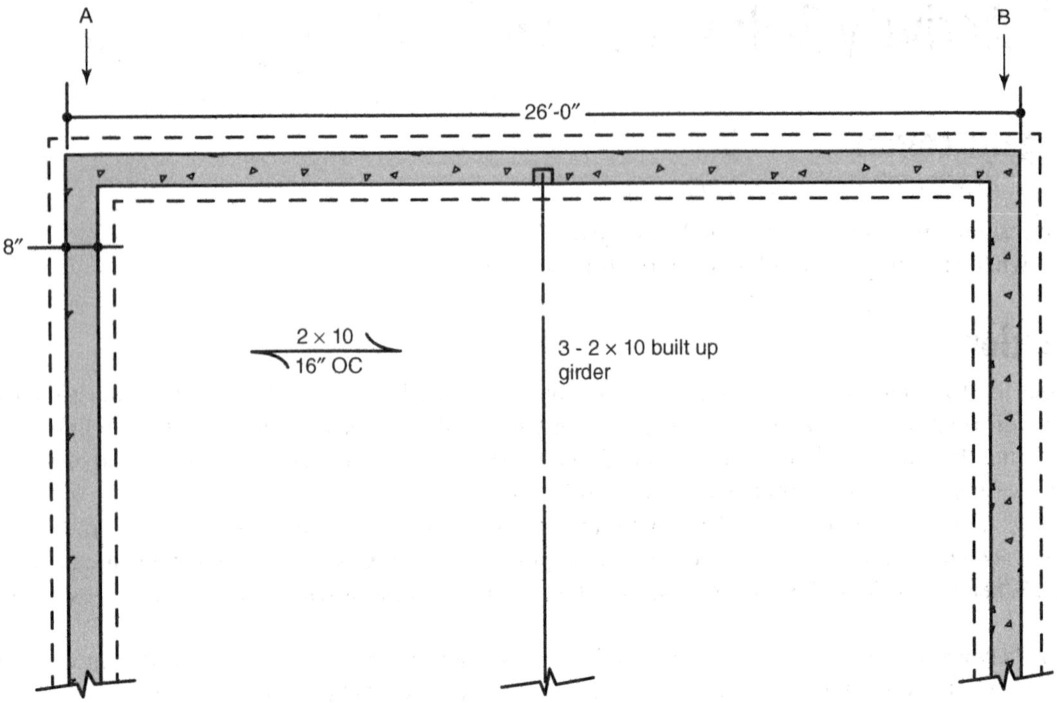

Goodheart-Willcox Publisher

Figure 1

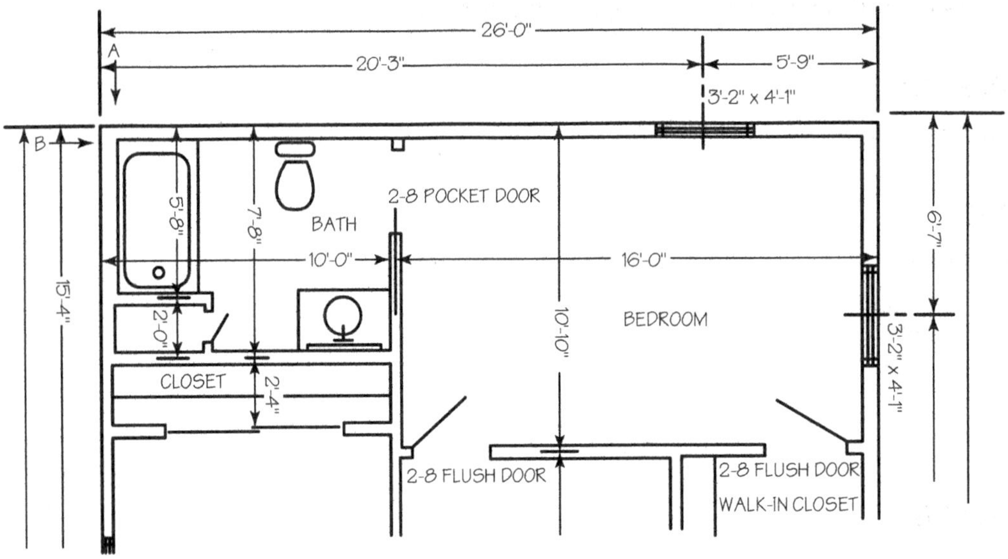

Goodheart-Willcox Publisher

Figure 2

Floor Layout

2. Set the 2″ × 4″ × 12′ on sawhorses for easy access as the layout is done. The floor system band joists will be laid out first.

3. Framing is spaced so that the sheet goods used for sheathing will always have a nailing surface at the ends or edges of the sheet. These sheet goods—plywood or oriented strand board (OSB)—are 4′ wide by 8′ long. Layout locations factor into overall dimensions. The most common layout spacings used are 16″ or 24″ on center (OC) since they are factors of both 48″ and 96″.

Name ____

4. Review the prints carefully and note any items that may affect the layout including the following:
   - Location of beams supporting the floor system.
   - Walls above which run parallel to the floor joists.
   - Plumbing locations, especially a toilet.
   - Openings that need to be built into the floor system.
5. Band Joist A is part of the floor system with a beam in the center. When laying out joists that will overlap over the beam, only one layout mark is measured out and then the floor joists are placed on either side of the mark for each side of the building. **Figure 3** indicates how this layout will look.

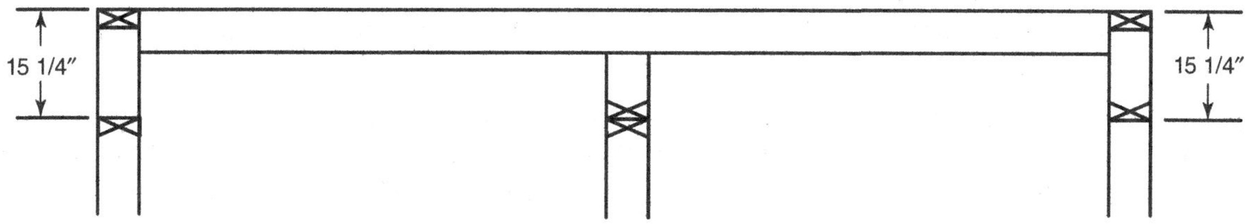

Goodheart-Willcox Publisher

Figure 3

6. Starting at the corner with the A arrow, what things in the floor plan above will affect this layout? The toilet and bathtub are on this side of the building, therefore the layout must not include a joist below these fixtures to interfere with the pipes. The walls creating the closet are parallel to the joists. Anytime there is a wall above that runs parallel to the joists, an extra joist is added to strengthen the floor system to support that weight above.
7. Start the layout on the 2×4 by placing the tape measure down the length of the board. The first mark is at the end of the board where the end band joist will be fastened to form the corner.
8. The floor joist symbol shows that the joists are to be placed 16″ OC. To make sure the end of the sheathing will land at the center of the floor joist, the mark will not be at 16″. The thickness of the joist is 1 1/2″; therefore, the center is 3/4″ from each edge. The mark placed on the band joist will be 3/4″ back from the 16″ at 15 1/4″ so the carpenter can see the line while fastening the joists in place. Use the square to draw a line across the board and place an *X* under the 16″ mark, forward of the line to show where the joist will be fastened next to the line.
9. To mark out the remaining joist locations, go to each consecutive 16″ mark and subtract 3/4″ to place the mark on the board. For example, the next 16″ multiple will be 32″, so the mark will be placed at 31 1/4″. Draw the line across the wood and draw an *X* on the forward side of the line, **Figure 4**.

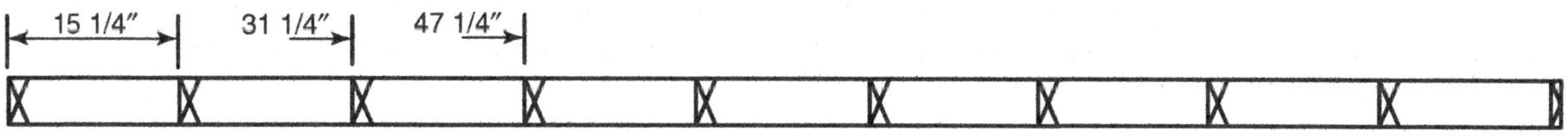

Goodheart-Willcox Publisher

Figure 4

10. Next, locate the fixtures and decide if any joists need to be moved. The toilet is against a 5 1/2″ wall with 1/2″ drywall on it, and the standard rough-in dimension to the center of the toilet waste line is 12″ from the wall. Add these measurements to get 18″ from the end of the building to the center of the pipe under the toilet. Measure and mark this on the layout board. There will be a 3″ pipe centered here. Will that pipe hit the joist at the 16″ mark?
11. Measure 1 1/2″ forward from the layout line and make a small mark. Then measure back 1 1/2″ from the pipe center (one half the width of the 3″ pipe). They overlap by 1/4″ showing that the joist will be in the way of the pipe placement. To solve this, place the joist on the other side of the layout line to make room for the pipe. Erase the *X* on the forward side of the layout line and re-mark it on the rear side of the line away from the pipe location, **Figure 5**.

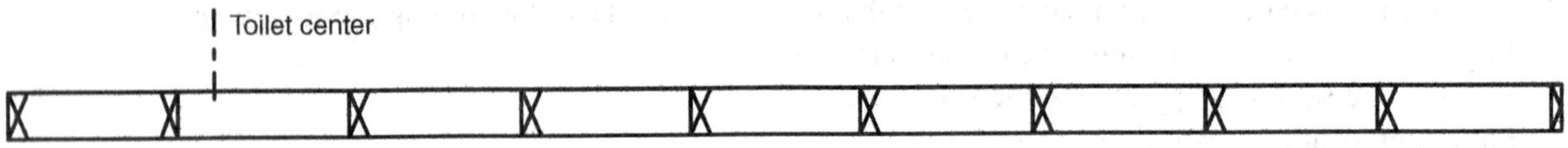

Goodheart-Willcox Publisher

Figure 5

12. The next fixture to check is the tub. The drain of the tub will have a trap below the floor next to the end wall by the drain. The dimension to the end wall is 5′-8″ or 68″ from the start of the floor system. In the layout, there is a joist at 63 1/4″ and another at 79 1/4″. The location of the tub drain will be between those two joists, so no adjustment is necessary.
13. The short wall at the end of the tub only goes a short distance across the room. Even though it is parallel with the joists, this length would not require adding the double floor joist below it. The walls to create the closet do run almost the full length of the joist therefore the joists here should be doubled to carry that extra weight.
14. By adding the dimensions of the tub and the linen closet, the first wall from the end of the layout is centered at 7′-8″ or 92″. The closest joist in the layout is at 95 1/4″. If a second joist is added to the rear side of this joist, it will be just under the wall to add the extra support needed.
15. Adding the dimensions of the bathroom and the closet, the second wall forming the closet is calculated at 10′ or 120″. There is a joist at 111 1/4″ and 127 1/4″ which places that wall above the center of the space between the joists. In this case the supporting joist can be added right under the wall at 120″. The final layout of the band joist for Wall A is shown in **Figure 6**.

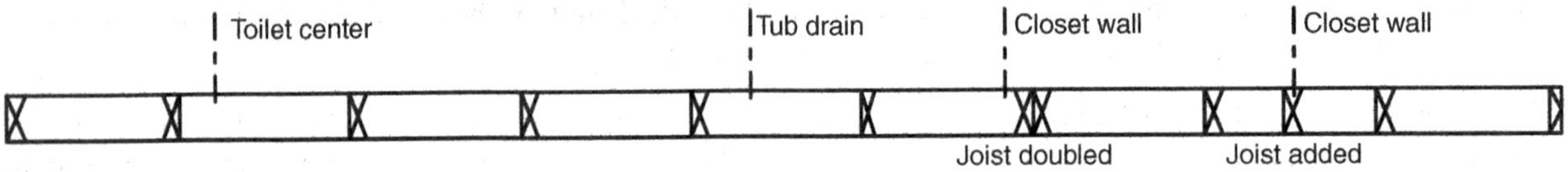

Goodheart-Willcox Publisher

Figure 6

16. Cut the end of the band joist at the center of a joist. The next band joist will start at the same center point. This allows both band joists to be fastened to that joist for strength.
17. Complete the layout for Band Joist B on the other side of the 2×4. Look for items that will need special consideration. The only thing on this side is the bedroom wall running parallel to the floor joists. Remember, the *X* marks are on the other side of the layout lines so the joists overlap over the beam. The completed layout should look like **Figure 7**.

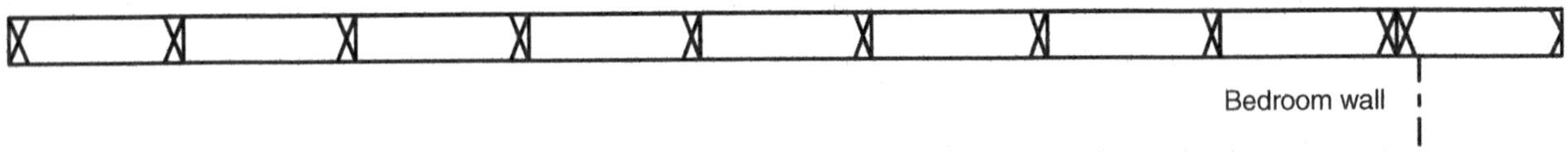

Goodheart-Willcox Publisher

Figure 7

Wall Layout

18. When laying out walls, the load bearing walls are prepared first and then the other walls will be fit in between them. The layout must create a continuous load path from rafters to the floor joists; therefore, the OC layout must match the floor joist layout on each side of the building.

Name ______________________________

19. On a 2″ × 4″ × 12′ wall plate, lay out the on-center studs first for Wall A. The walls will typically have the same 16″ OC layout to ensure the sheathing has good nailing on all edges and ends. Since Wall A is a loadbearing wall, it will start at the outside edge of the floor system and the studs should line up with the floor joists below. The first layout mark should be at 15 1/4″ just like the band joist was started. Sometimes a carpenter will tack a nail at this mark and hook the tape measure to it and mark right at the 16″ marks down the board instead of having to remember to subtract the 3/4″ at each mark, **Figure 8**.

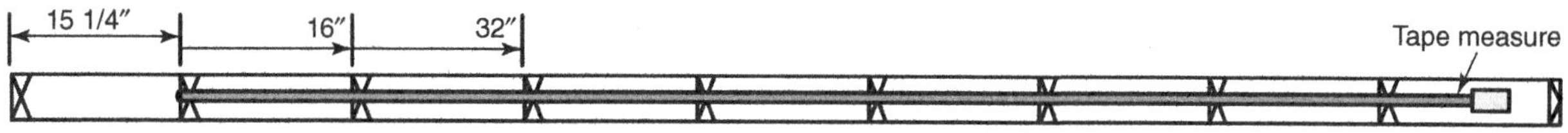

Goodheart-Willcox Publisher

Figure 8

20. The corner framing needs to be decided first. A corner must provide nailing surfaces inside for the drywall that will be on each wall and sheathing on the outside. For energy efficiency, the least amount of lumber should be used to allow insulation to be installed. A two-piece stud corner will meet those requirements. Lay out the corner as shown in **Figure 9**.

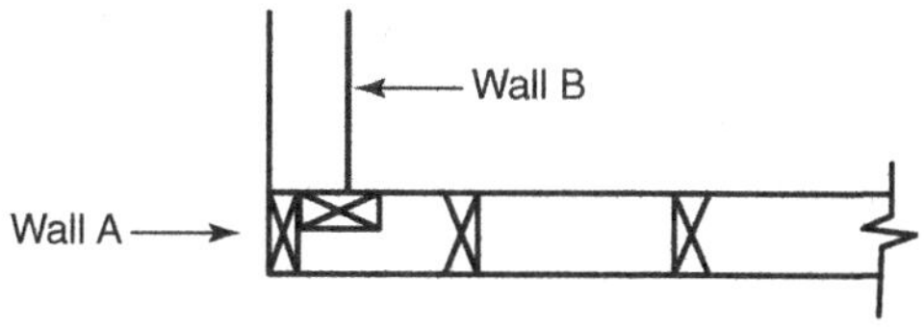

Goodheart-Willcox Publisher

Figure 9

21. Recall that the first joist mark in the floor system was moved to the other side of the layout mark to accommodate the plumbing for the toilet. Do the same at the first stud layout line to ensure the continuous load path of the framing members.

22. The intersecting walls further down the wall will also need intersection framing to provide nailing surfaces for the interior finishes. A two-piece corner or a ladder type corner could be used for these. Measure and mark the center of each wall to decide which type fits the best.
    - The center of the first wall at the end of the tub falls in the middle of two stud layouts. The corner ladder will work best here.
    - The center of the second wall falls right next to OC stud layout so a two-piece corner will work here.
    - The third wall falls between two stud layouts, so another ladder corner will work here. **Figure 10** shows what the final layout should look like.

Goodheart-Willcox Publisher

Figure 10

23. On the other side of the 2×4, lay out Wall B. The first stud mark will be the one to complete the corner framing to provide nailing for the interior finishes.

24. This wall is non-loadbearing since it runs parallel to the floor joists. The studs do not have to line up with any other framing for a load path. The sheathing will determine the layout of the studs and will start at the outside of the corner on the other side of Wall A.

25. Overhang the tape measure 6″ to equal the width of the 2×6 wall plus the sheathing on Wall A. Place the first mark at 15 1/4″ just like the floor layout so the piece of sheathing will end on the center of a stud. Start the layout as shown in **Figure 11**.

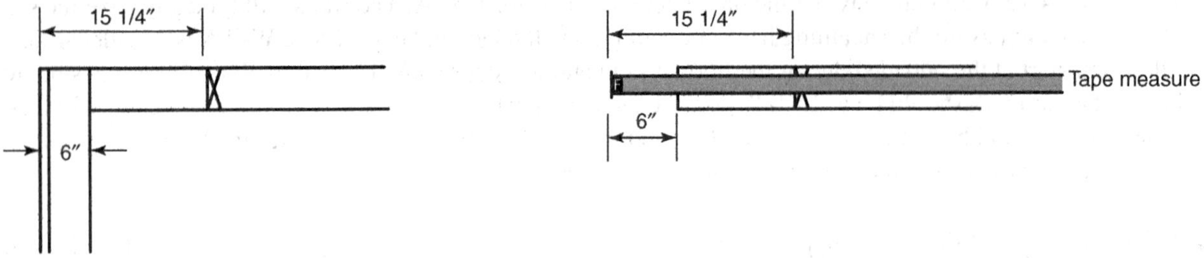

Goodheart-Willcox Publisher

Figure 11

26. Tack a nail at this mark and stretch out the tape measure to mark the 16″ OC marks down the length of the wall plate.
27. Looking for special framing issues, this wall has only the wall intersection at the bathroom door to add framing for interior nailing. The location of the wall falls in the middle of two studs which would best be done with a ladder intersection. The completed layout is seen in **Figure 12**.

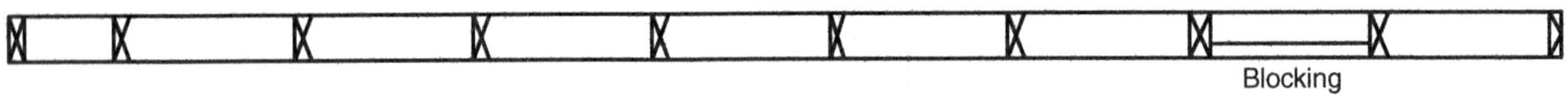

Goodheart-Willcox Publisher

Figure 12

Review and Analysis

1. Describe why it is important that the floor joists and the wall studs line up from bottom to top of the structure when doing the layouts?

____

____

____

2. When a corner or wall intersection is framed, what must be provided to prepare for future stages of construction?

____

____

____

3. What information is given with the floor joist symbol?

____

____

____

Name ______________________ Date ____________ Class ________

Activity 3-4: Project—Drawing the Floor Plan for the Corner Unit

Learning Outcomes

After completing this lab activity, you will be able to:

- Draw a floor plan of a corner unit.
- Mark the proper spacing for studs in a wall.
- Calculate materials needed to frame a wall.

Introduction

The corner unit will be used as an introduction to framing, wiring, plumbing, and root framing. It can also be used for interior and exterior finishes, such as drywall, painting, and exterior coverings.

As you progress through the textbook and learn more about the phases of construction, you will apply this knowledge to this corner model. The first step will be to draw and dimension this building project. You can estimate lumber for a project by calculating one stud per foot of wall, one stud for each corner, two more studs for each window and door, and lumber for top, bottom, and cap plates.

Text Reference

Chapter 3, *Framing Systems*

Equipment

The following tools and equipment are needed to perform these activities:

- Architect's scale
- Graph paper with 1/4″ grid or lined paper
- Mechanical pencil
- Ruler or drafting triangle

Procedure

When building this project, you will need to use 2×6s for your long wall, which will house the plumbing. The smaller side walls will be constructed with 2×4s, like interior walls in a home. The overall height of this corner unit will be 36″. This measurement includes the bottom, top, and double-top plates. The placement of studs in the side walls must match so that ceiling joists will lay straight and allow plumbing and electrical to be roughed in.

Eventually, electrical power will be provided to allow the fixtures and outlets to be checked for functionality. This floor plan will fit on a single piece of paper in the landscape position. The drawing will be drawn to a scale of 1 1/2″ = 1′-0″. This translates to 1/4″ = 2 real inches, or six squares of the graph paper equals 1′. Use the drawing **Figure 1** for reference.

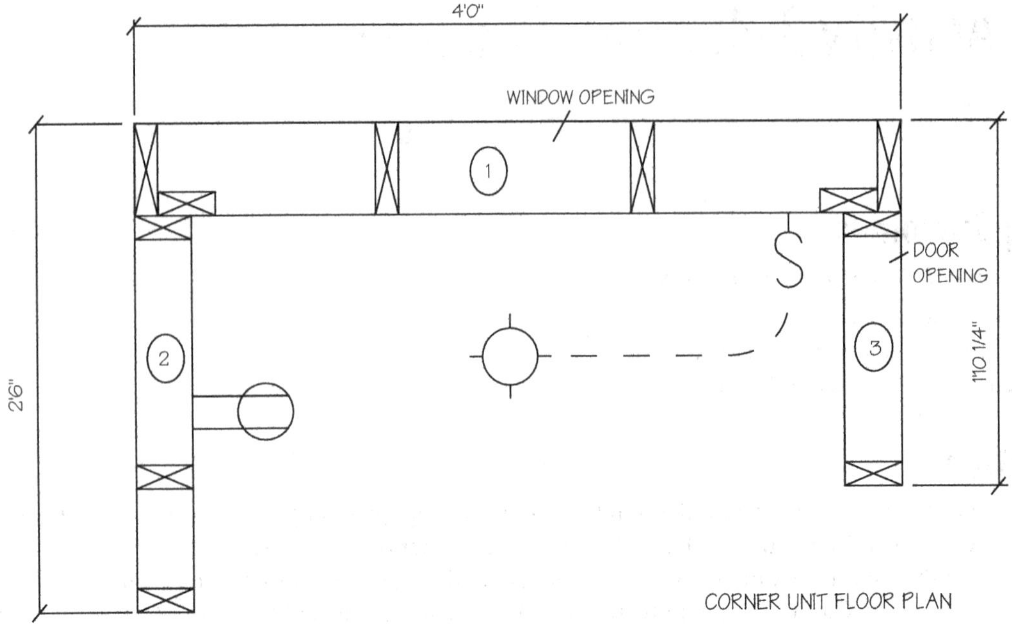

Goodheart-Willcox Publisher

Figure 1

Construction Lines for Wall Layout

1. Lay your paper in the landscape position.
2. Find the center of the width of your paper. Using the architect's scale, measure 2′-0″ to the left and right of the center.
3. From these measurements, draw two vertical construction lines that you will darken later.
4. Find the center of the height of your paper and measure 1′-6″ up and from the center 1′-6″ down. Draw your horizontal construction lines.

> **NOTE**
> From these construction lines, you will draw your floor plan.

5. From the top construction line, measure down 5 1/2″ and draw a line parallel to the line you just measured from. This will be your plumbing wall.
6. From your outside vertical lines on the left and right, measure in 3 1/2″ and draw the inside of the 2×4 framing for the end walls of the unit.
7. To create the left wall, measure 2′-6″ from the outside of the plumbing wall and mark where the left wall will end.
8. To create the right wall, measure 1′-10 1/4″ from the outside of the plumbing wall and mark where the right wall will end.

Marking Stud layout

1. Starting from the left side of the plumbing wall measure 1 1/2″ in and draw the vertical line that creates the 2×6 stud. Draw an *X* in this rectangle to show that this is a stud.
2. The next measurement will be 15 1/4″ from the left corner. The easy way to draw this is to lay it out just like you would lay out stud placement on top and bottom plates.

> **NOTE**
> Studs are laid out 16″ OC (on center) but the first stud needs to be laid out at 15 1/4″ so the sheathing falls on the center of a stud when it is installed.

Name ____

3. From the left side of the plumbing wall measure 16″ to the right. Then from that mark, subtract 3/4″ (mark it) and then go back to the 16″ mark and add 3/4″ and mark it. At the 3/4″ marks draw your vertical lines and draw your *X*. Every time you locate a stud, put an *X* in it.

> **NOTE**
>
> When laying out an actual wall plate, note the red highlighted measurements on a tape measure. These highlights are located on the 16″ layout measurements along the tape for fast and easy layout marking. To lay out your studs, mark 3/4″ to one side of each 16″ marking and draw lines with a speed square. Draw an *X* on the side of the mark where the stud will be fastened.

4. Repeat this process to locate the rest of the 2×6 studs in your plumbing wall.
5. The plumbing wall needs two corners. Draw in a 2×4 that lies flat to create an L-shape, one left hand and one right hand corner.
6. Repeat this process to draw in the 2×4 studs on the side walls. Make sure the first two studs line up with each other across the unit. This is important for ceiling and roof layout so there is a continuous path of load bearing from the foundation to the roof.
7. Darken all your object lines.

Adding Electrical and Dimension

1. Add your electrical symbols, duplex outlet, wall switch, and light. Use the dashed hidden line to show the wiring path.
2. Dimension your drawing.
3. Number your walls as shown in **Figure 1**. This will be the order you build your walls.

Review and Analysis

1. When roughing out a drawing, what type of line do you draw first?

____

____

2. What is the standard spacing of studs in a wall? What is the first measurement needed for marking the location of the first stud on a wall?

____

____

3. How much lumber would be required if this model were built full size?

____

____

____

Notes

Name ______________________ Date ______________ Class __________

Activity 3-5: Project—Building the Corner Unit

Learning Outcomes

After completing this lab activity, you will be able to:

- Lay out studs for a wall with an outside corner.
- Mark the proper spacing for studs in a wall.

Introduction

With the completed plans in hand, it is time to build the corner unit that will serve as a practice project while exploring the residential construction industry. This corner unit will provide a model structure to practice skills needed during wall framing, roof framing, and some of the mechanical systems found in a home.

Throughout the build process, refer to the corner unit floor plan, **Figure 1**.

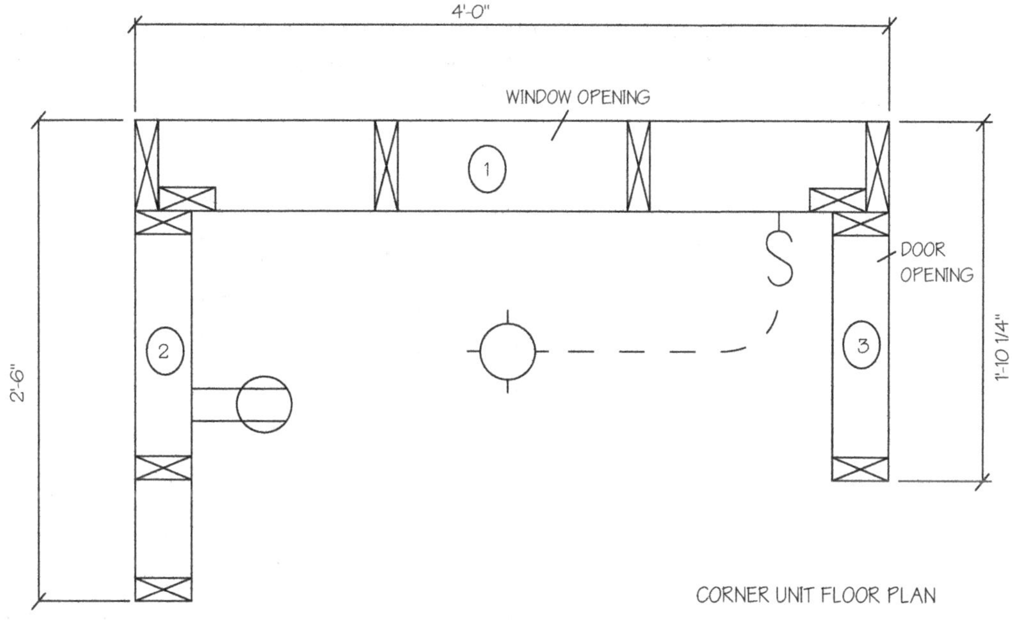

Goodheart-Willcox Publisher

Figure 1

Text Reference

Chapter 3, *Framing Systems*

Equipment

The following tools and equipment are needed to perform these activities:

- 16d sinkers or 16d common nails (60)
- 1×4×6′ lumber (1)
- 2×4×8′ lumber (4)
- 2×6×8′ lumber (3)
- 8d common nails (30)
- Carpenter's pencil
- Framing hammer
- Framing square
- Level, 2′
- Measuring tape
- Power miter saw or circular saw
- Quick-acting clamps (F-clamps) (2)
- Safety glasses
- Speed square

Safety Review

Follow all safety procedures specified by your instructor and specified in any appropriate manufacturer's information. The following are some additional safety-related items to keep in mind:

- Wear safety glasses at all times.
- Keep your fingers 6″ away from all blades when cutting.
- Measure twice, cut once.
- Clamp lumber down when needed.

Procedure

The plumbing wall will use 2×6s and side walls will use 2×4s. Remember that this lumber is 1 1/2″ thick when figuring out the lengths for your wall studs. The wall studs are the height of the wall minus the thickness of the top, bottom, and double top plate. Each stud should be nailed to the top and bottom plate with two 16d nails evenly placed through the face of the plate into the end of the stud.

1. Crosscut 2×6 lumber for top and bottom plates at 48″ for the plumbing wall.
2. Lay both plates on a sawhorse next to each other and line them up end to end. Mark the layout locations on one of them using the crow's foot mark, **Figure 2**. To save time when marking layout marks, use the framing square to mark the lines across both pieces at once.

Goodheart-Willcox Publisher

Figure 2

3. First, lay out the end studs forming the corners. Measure in from the ends 1 1/2″ and draw the lines across both pieces. Draw an *X* in the end space to show the stud location. Then lay out the size of the 2×4 along the edge to form the corner post.
4. Lay out stud markings every 16″. The first mark is at 15 1/4″ from the end of the plate to make sure the wall sheathing will end on the center of a stud to reduce waste. The next layout marks are 16″ from the first. After drawing the line across both pieces, place an *X* next to the line on the side where the stud will be fastened.
5. Crosscut your 2×6 lumber to length for the four wall studs. If the completed wall is going to be 36″, the thickness of the plates needs to be deducted to calculate the length of the studs. There will be three plates to complete the wall. Each is 1 1/2″ thick, for a total of 4 1/2″ to be subtracted from the overall 36″ height. The studs will be 31 1/2″.
6. Crosscut your 2×4 lumber to length for the two wall studs needed to build the corners.
7. Construct the two corners before nailing them into the wall. Lay the 2×4 on the floor on edge and place the 2×6 on top of it, lining up the ends and the edge. Nail through the 2×6 into the 2×4. The first and last nail should be placed 4″ from the end. The last nail should be centered between the first two.

NOTE

The three stud corners being built are designed to provide an inside nailer for the interior drywall without any extra lumber filling the corner, therefore allowing more room for insulation to go completely out to the outside of the corner. This is sometimes referred to as a California corner.

8. Assemble your first wall as shown in the floor plan. Lay the plates down on the floor on edge with the layout marks facing the center between the two. Start by laying out the two corners, lining up the corners in the correct orientation to match the plan. Then, lay the two studs in their position.
9. Start nailing at one corner. Place a knee on the corner stud, putting pressure on it and holding it in place while nailing. From the bottom of the plate, drive two 16d nails through the plate into the stud. Add one more nail into the 2×4 used to make the corner post.
10. Repeat this process for each of the studs across the plate. Be sure to line up the side of each stud with the layout mark and over the *X*. Also, make sure the edge of the stud is flush with the edge of the plate.
11. Repeat this process for the other plate on the other end of the studs. When complete, stand the wall up and hammer each of the nails at each stud to ensure the plates are tight to all the studs.

NOTE

Focus on hitting the nail with the center of the hammer face. Hold the hammer toward the end of the handle. Holding the hammer too closely to the head will result in a weaker blow. If you hit the nail lightly and repeatedly, the nail may bend, and you will need to pull the bent nail out. If the nails are not easily removed, put a wood block under the hammer for more leverage, **Figure 3**.

Goodheart-Willcox Publisher

Figure 3

12. Repeat these steps to cut and assemble the two side walls with the 2×4 lumber to match the plan drawing.

NOTE

When assembling the corner unit, make sure stud placement on the side walls match.

13. Stand the 2×6 wall up and use the level to check an end to see if it is plumb. If it is not, stand on the bottom plate on the floor and push the top plate in the direction needed to make the end plumb. Once plumb, clamp one of the side walls in position on the corner post. Start a couple of nails in the 2×4 side stud to nail the stud to the corner post but do not drive them into the corner stud.
14. Check the unit for plumb along the short wall now. Adjust the intersection with the clamp until it is plumb and then attach the side walls to the plumbing wall by driving the nails in the rest of the way. Use the same nail pattern as when building your corners.

15. Repeat these steps for the other end wall. Before final nailing again, check for plumb on both sides of the corner post.

16. Add the double top plate to the wall construction to tie all the intersecting walls together and add some rigidity to the overall frame created. Install the plates to overlap the adjacent wall to add this strength, **Figure 4**. Add the double top plate using 8d common nails. Place nails over existing studs to avoid nails interfering with plumbing and electrical rough-in that might be installed between the studs later during the mechanical rough-in stages. Do not nail over the 16d nails; nail next to where the 16d nails are located.

Goodheart-Willcox Publisher

Figure 4

Squaring the Corner Unit

Each of the corners of the unit need to be square with each other just like when building a home. This can be accomplished in a couple of ways. Cut the 1×4 into one piece 42″ and one 24″ long. See **Figure 5** to see how this should appear when complete.

Goodheart-Willcox Publisher

Figure 5

17. One method is to use the 3-4-5 triangle method to square the corners. A right triangle with the two legs of 3 and 4 units and the hypotenuse 5 units will always create a square corner. The larger the triangle measured, the more accurate this method is when squaring a corner, so any multiple applied to each side will give an accurate corner. The length of the longer end wall is 30″ and is equal to 3 times 10. That will be the 3 leg of the triangle. If we use 10 as the multiplier for the other two sides of the triangle, we will have a good fit for the unit being built. Along the 2×6 wall from the corner by the long end wall, make a mark at 40″ which is the 4 leg of the triangle times 10. Attach the 42″ piece of 1×4 diagonal to the 2×4 top plate about 22″ from the corner. Measure the hypotenuse (the 5 leg) by hooking the tape measure on the end of the 2×4 wall and holding it diagonally to line up with the 40″ mark on the 2×6 plate. Adjust the position of the end wall until the 50″ on the tape intersects with the 40″ mark on the 2×6 and then nail the other end of your diagonal to the 2×6 top plate. This method is used frequently by builders to make sure corners of any size are square.
18. The other method is to use a framing square to square the corners. This will work for this small model being built but is not accurate for a large house project. Use it to square up the short end wall. Use an 8d nail to attach the 24″ piece of 1×4 lumber to the 2×6 plate about 16″ from the corner. Hold the square with the body of the square along the top plate of the 2×6 wall and the tongue out along the end wall. While holding the square tight to the 2×6 plate and tight to the inside of the corner, adjust the end wall so it is tight to the square and then nail the other end of the piece of 1×4 on the end wall plate. The corner unit is ready for the next stages of the construction path.

Review and Analysis

1. Why do double top plates overlap in the corners?

____

____

____

2. Demonstrate and explain how to pull a nail out that was nailed in incorrectly or bent and could not be used.

____

____

____

3. Why should the 3-4-5 triangle method be used to square a wall when building a house rather than using a framing square?

____

____

____

Notes

Name ______________________ Date ______________ Class __________

Activity 3-6: Skill Building—Framing a Door and Window Opening

Learning Outcomes

After completing this lab activity, you will be able to:

- Identify the parts needed to frame a door opening.
- Demonstrate how to frame a door opening.
- Identify the parts needed to frame a window opening.
- Demonstrate how to frame a window opening.

Introduction

In this activity you will frame a door opening in a small wall. Then you will deconstruct the wall and reuse the parts to frame a window opening. You will be using standard wall framing lumber for your geographic area. Traditional door and window heights are framed at 6′-10″ in height. This measurement will be used for the door and window framing. You will need to learn the parts of a wall. Use **Figure 1** for reference throughout this activity.

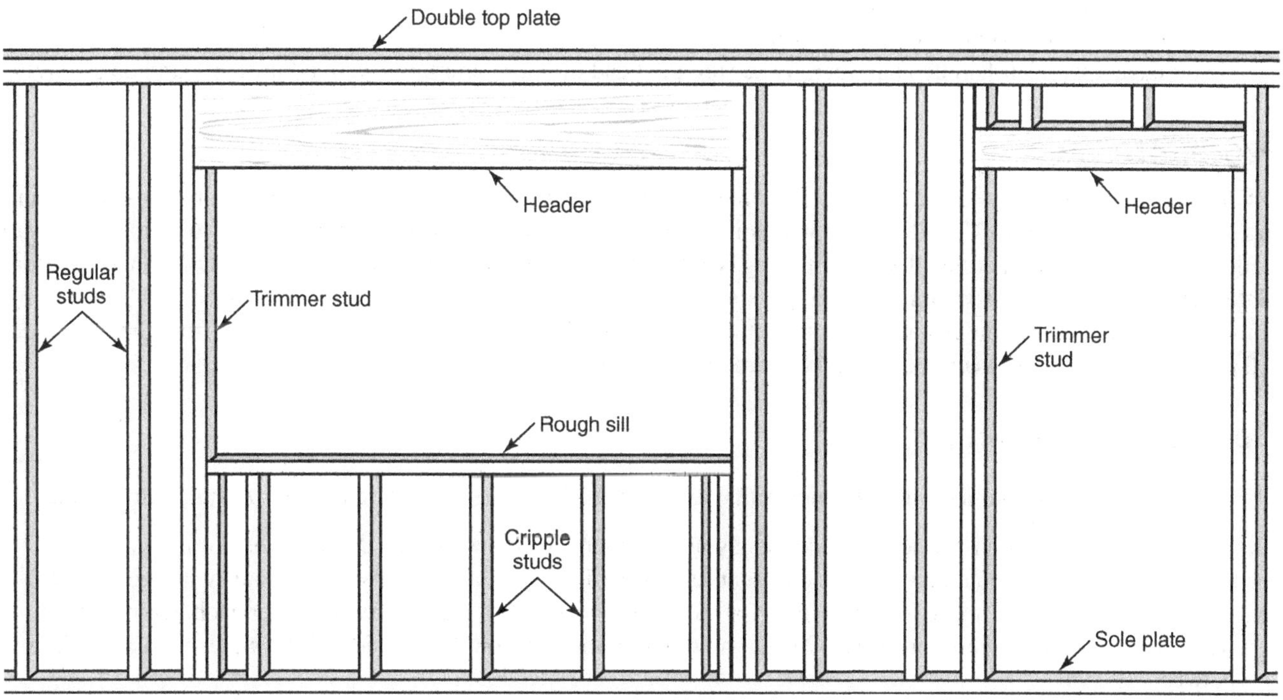

Goodheart-Willcox Publisher

Figure 1

Text Reference

Chapter 3, *Framing Systems*

Equipment

The following tools and equipment are needed to perform these activities:

- (1) 1/2″ × 3 1/2″ × 36″ plywood (spacer)
- (1) 2×4×8′ lumber
- (7) 2×4×92 5/8″ precut studs
- 80 grit sandpaper or light-colored paint
- Box 8d common nails or green vinyl sinkers
- Box 16d common nails or green vinyl sinkers
- Carpenter's pencil
- Circular saw or power miter saw
- Measuring tape
- Roofing nails (10), 1″ to 1 1/2″ long
- Safety glasses
- Speed square

Safety Review

Follow all safety procedures specified by your instructor and specified in any appropriate manufacturer's information. The following are some additional safety-related items to keep in mind:

- Wear safety glasses at all times.
- Wear steel-toe shoes on the jobsite.
- Be aware of others around you as you swing your hammer back, before swinging forward.
- If you have a wooden-handled hammer, make sure the head of your hammer is wedged properly and on securely.

Procedure

Doors are generally located in a corner of a room to not take up very much wall space. We will frame this door opening in the corner of our 4′-0″ wall that we will build. The door opening will be a standard door, 2′-6″ wide. This wall will have one outside corner that the door opening will be next to. Each 2×4 stud will be nailed into the top and bottom plates with two 16d nails. Toenailing for cripple studs will use 8d nails. Door jambs are 3/4″ thick and a 1/2″ space for shims will be allowed for the left and right side of the door jamb. Top and bottom plates will be cut from the 8′-0″ stud. All other lumber used between the top and bottom plates will use the 92 5/8″ precut studs.

Door Rough Opening

1. Crosscut the 2×4×8′ lumber into two pieces that are 4′ long. These pieces will be for the top and bottom plates.
2. Set up the 2×4 plates on sawhorses to do the layout together just like what was done for the corner unit construction. On one plate, layout the studs marking every 16″. Then mark across both plates with the square. On the left end, layout space for three studs for the corner post with 1 1/2″ spaced lines for each. Do not mark studs with an *X* yet, except for the two *X*s that mark the corner post on the left side of the plates with a block in between the two.
3. Locate the center of the door opening on your top and bottom plates as shown on the wall layout plan, **Figure 2** and **Figure 3.**

Name ______________________________

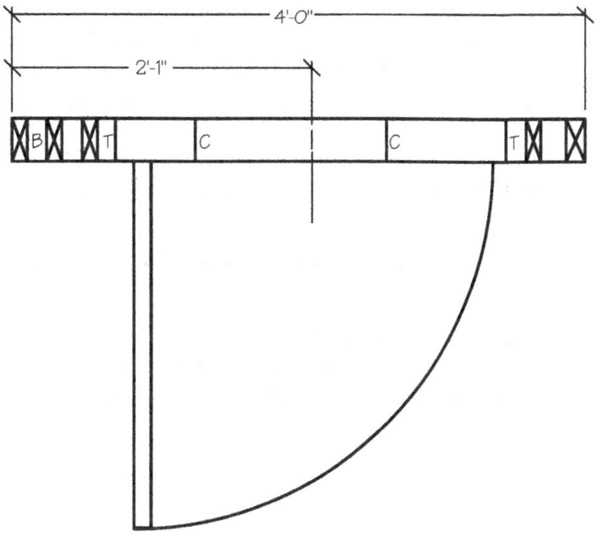

Goodheart-Willcox Publisher

Figure 2

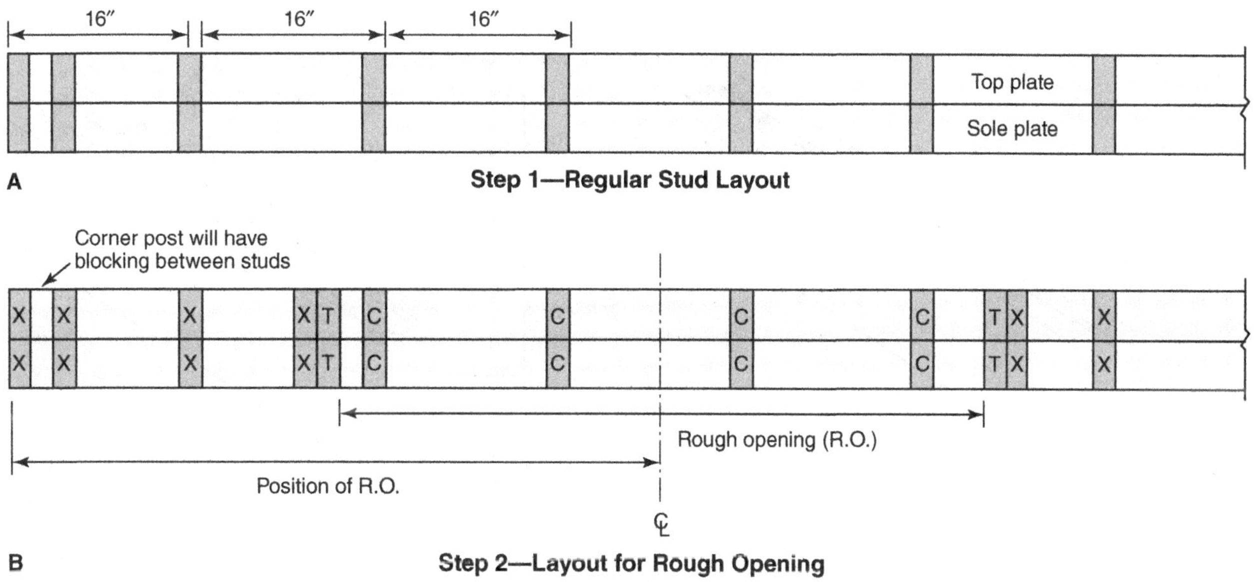

Goodheart-Willcox Publisher

Figure 3

4. Measure and place a mark at 15″ (1/2 of the door width) to the left and right of the door center mark.
5. To the outside of these marks, measure and mark 1 1/4″. This marks the space needed for the 3/4″ door jambs and 1/2″ shim space.
6. Add 1 1/2″ to the left and right of those marks to show the trimmer studs. Mark this space with a *T* to denote the placement of the trimmers. These are sometimes called Jack studs.

> **NOTE**
> To the outside of these marks, measure 1 1/2″ for the first full length studs next to the opening. These are sometimes called the king studs.

7. Mark these last stud markings with an *X* because they are full length. See **Figure 3.**
8. On the top plate only, in the stud markings between the trimmer studs, mark them with a *C* for *cripple stud.*

9. To make the header, first measure the distance between the full stud layouts on either side of the opening. Crosscut two pieces of 2×4 to fit between these studs. Cut the piece of 1/2″ plywood to the same length. This will be for the header, which is sometimes called a lintel. The header carries the weight of the structure above and must be supported by trimmer studs going all the way down to the bottom plate.

10. Stack up the header pieces with the 1/2″ plywood between the two 2×4s. Line them up end for end and side to side and nail the header pieces together with 16d nails, three sets of two nails placed on each end and then one in the center. See **Figure 4**.

> **NOTE**
>
> A shortcut for figuring out header length for a door opening is to take the door width and add 5 1/2″ to determine the length of the header. This is the door width plus 1 1/2″ × 2 for the trimmer studs, plus 3/4″ × 2 for the door jamb thickness, plus 1/2″ × 2 for the shim space. The door header fits between the king studs and is above the trimmer studs.
> 1 1/2 × 2 = 3
> 3/4 × 2 = 1 1/2
> 1/2 × 2 =1
> Add it together: 3 + 1 1/2 = 4 1/2, 4 1/2 + 1 = 5 1/2
> This gives you 5 1/2″ to add to the door width for the length of the header.

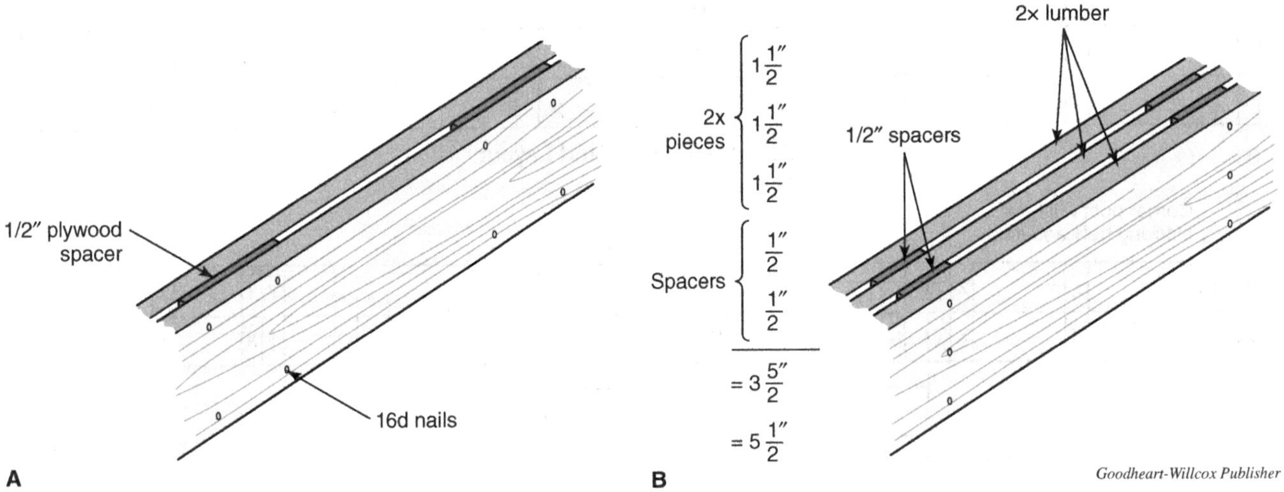

Goodheart-Willcox Publisher

Figure 4

11. Crosscut two pieces of 2×4 to the length of the trimmer studs. The header height off the floor will be 6′-10″. The thickness of the bottom plate needs to be subtracted from that height to get the length of the trimmer.

6′-10″ – 1 1/2″ = 6′-8 1/2″

12. Save the cut-offs left from the trimmer studs for the cripple studs over the door header.

13. Nail the trimmer studs to the king studs with 16d nails every 12″. Before nailing, be sure the trimmer and king studs are lined up at the bottom end.

14. Place the top plate on the header with the ends of the header in line with the king stud layouts and copy the locations of your cripple studs to the header.

15. Place the header between the king studs on top of the trimmer studs. Nail together with four 16d nails through the king stud into the header. Place the nails evenly.

16. Cut three blocks of 2×4 at 12″. Build the outside corner with two 2×4s with these blocks in between them. Place one 2×4 on the floor and place one block on each end and one in the center. Place the second 2×4 on top of the blocks. Line each up carefully and nail with three 16d nails in each block. Roll the whole corner post over and nail two 16d nails in each block from the other side.

17. Nail the corner into the left side of the top and bottom plates.

18. Nail the door frame in place, matching the king and trimmer studs to the marks on the top and bottom plates.

19. Nail the last 92 5/8″ stud at the end of the plates.

20. At one end of the header, measure the distance between the header and the top plate. Crosscut the cripple studs to this length to fit above the header.

21. Nail each cripple stud in with two 16d nails through the top plate and toenail four 8d nails in the sides of each stud into the header. See **Figure 5**.

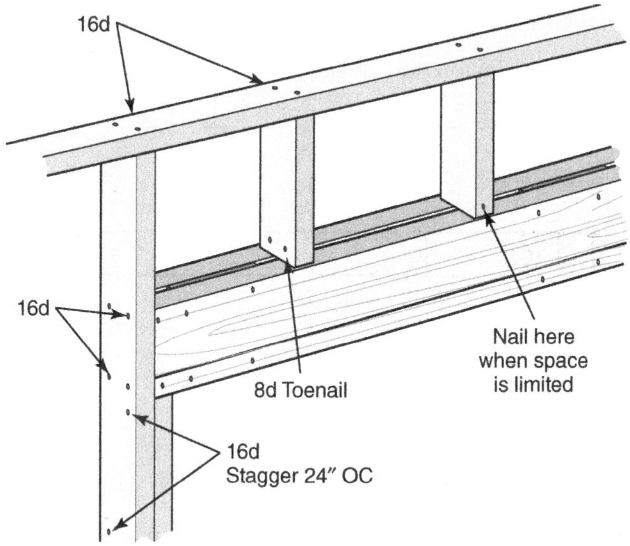

Goodheart-Willcox Publisher

Figure 5

NOTE

Toenailing is when a nail is driven at an angle from the side of a stud into the intersecting framing member. To help with toenailing, use roofing nails.

22. The cripple studs will want to move as you toenail them in. Use the roofing nails as stops, **Figure 6**. Nail the roofing nails in sideways to be even with your cripple stud marks. Toenail from the other side first. Then toenail the remaining side. You can leave the roofing nails in place or take them out.

Goodheart-Willcox Publisher

Figure 6

Window Rough Opening

Disassemble the 4′-0″ wall with the rough door opening. Do not take apart the outside corner, trimmer and king stud assemblies, or the header. Save all the other parts and take out the nails so they can be reused for the window rough opening to be framed next. The window will be 24″ × 36″ with a 1/2″ of shim space all around the window frame to allow the squaring of the window. We will place the window in the center of this 4′-0″ wall.

23. Turn the top and bottom plates over and mark for studs every 16″ on center (OC). An option is to sand or paint over the existing stud markings to start fresh.
24. Locate the center of the window opening on the top and bottom plates, **Figure 7.**

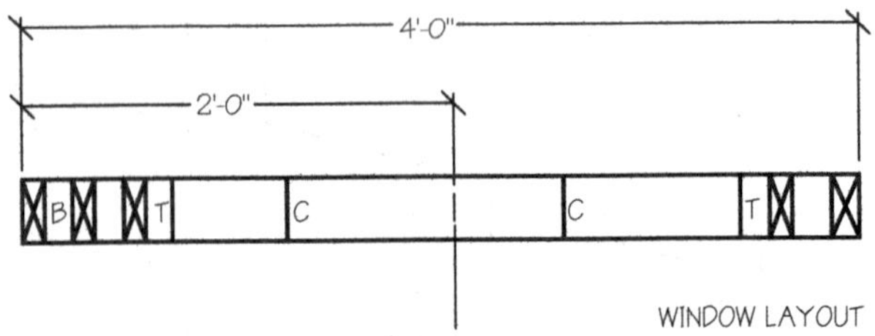

Goodheart-Willcox Publisher

Figure 7

25. Measure and mark 1/2 of the window width to the left and right of the window center mark.
26. To the outside of these marks, measure and mark 1/2″ to allow for window shims.
27. Add 1 1/2″ to the left and right of these marks to allow for trimmer studs.
28. To the outside of these marks, measure 1 1/2″ for the king studs.
29. Mark the stud spaces appropriately for the full-length studs, trimmers, and cripple studs, **Figure 7.**
30. Crosscut the header from the door opening to fit between the king studs for the window opening. Check to see where the nails are and do not cut through them.

NOTE

A shortcut for figuring out header length for a window opening is to take the window width and add 4″ to determine the length of the header. This is the window width plus 1 1/2″ × 2 for the trimmer studs, plus 1/2″ × 2 for the shim space on each side of the window.
1 1/2 × 2 = 3, plus 1/2 × 2 =1, will give you a total of 3 + 1 = 4″.
This results in 4″ to add to the window width for the length of the window header.

31. Put the header next to the top plate and copy where the cripple studs are located.
32. Nail the header to the king studs at the top of the trimmers. Remember, trimmer studs are already attached to your king studs.
33. Measure the space between the trimmer studs. It should be the width of window plus 1″.
34. Crosscut your rough sill to this measurement. This will be for the bottom of the window opening.
35. Place the rough sill next to the top or bottom plate and copy where the cripple studs are located, just like was done with the header.
36. Measure down from the header, the height of the window plus 1″ and mark it on both trimmer studs. This allows 1/2″ top and bottom for shims.
37. From that mark, measure down another 1 1/2″ for the rough sill thickness.
38. Measure from the bottom plate to this rough sill mark for the length of the cripple studs below the window. Cut two cripples at this length.

Name ____

39. Nail one cripple stud under the rough sill mark to each trimmer, lining it up with the bottom end. See **Figure 1**.
40. Nail the window assembly to the top and bottom plate matching each of stud markings.
41. Crosscut cripple studs to length for above the header and below the rough sill.
42. Nail them in place with 16d nails through the top and bottom plates, 8d nails for toenailing. Remember to use roofing nails to help hold the cripple studs in place when toenailing.
43. Nail in the corner assembly and the other full-length stud.

Review and Analysis

1. List the parts of a rough framed door assembly.

____

____

____

2. List the parts of a rough framed widow assembly.

____

____

____

3. What size header is needed for a double door 5′-0″ wide?

____

____

____

4. What is the purpose of a header in door and window openings?

____

____

____

Notes

Name ____________________ Date ____________ Class ____________

CHAPTER 4 Drying In

Activity 4-1: Chapter 4 Review

Answer the following questions after studying the chapter.

Lesson 4.1

1. _____ *True or False?* A house flipper will not see a profit on the home until it is sold. (4.1-1)
2. _____ The mission of the _____ is to educate and prepare individuals for construction careers. (4.1-1)
   A. Construction Management Association of America
   B. National Institute of Building Sciences
   C. American Institute of Constructors
   D. Home Builder's Institute
3. _____ *True or False?* To succeed in a career as a teacher or trainer, you must be able to effectively communicate your construction expertise. (4.1-1)
4. _____ To be successful in sales, it is *most* important to have strong _____ skills. (4.1-1)
   A. technical
   B. interpersonal
   C. leadership
   D. writing
5. _____ *True or False?* Inspectors have a more physically demanding job than most other building trade positions. (4.1-1)
6. _____ When damage occurs to an insured home, the insurance company sends a(n) _____ to survey the damage and determine the extent of the needed repairs. (4.1-1)
   A. insurance adjuster
   B. insurance agent
   C. building inspector
   D. policyholder

Lesson 4.2

1. _____ *True or False?* Builders are free to build to a higher standard than the building code. (4.2-1)
2. _____ *True or False?* Using best practices often saves the client money in the short term. (4.2-1)
3. _____ *True or False?* Meeting building code requirements is always the best practice, even if more effective products than the code requires are available. (4.2-1)
4. _____ Who determines where to use best practices? (4.2-1)
   A. Builder and client
   B. Client and architect
   C. Architect and inspector
   D. Builder and architect

Lesson 4.3

1. _____ When should electricians, plumbers, and HVAC technicians begin installing their systems in the home? (4.3-1)
   A. Before the building has been dried in
   B. During the drying-in process
   C. After the building has been dried in
   D. After the finishing work on the building has been completed

2. _____ *True or False?* When installed properly, a water-resistant barrier should be visible when the job is complete. (4.3-1)

3. _____ *True or False?* Windows and doors must be flashed into the water-resistant barrier system to keep water out of the framing. (4.3-1)

4. _____ Which material can be used for a home's water-resistant barrier? (4.3-1)
   A. Fiberglass insulation rolls
   B. Foamboard insulation
   C. Pine lumber
   D. Oak lumber

Lesson 4.4

1. _____ *True or False?* Felt paper is lighter and easier to use than synthetic underlayment. (4.4-1)

2. _____ Ice-and-water barriers are typically made of which material? (4.4-1)
   A. Rubberized asphalt compound
   B. Cellulose fibers impregnated with bitumen
   C. Aluminum sheathing
   D. Polypropylene sheets

3. _____ Which statement about slate, clay, and concrete roofing tiles is true? (4.4-1)
   A. They are among the least expensive roofing options.
   B. They are light and easy to handle.
   C. They offer good fire resistance.
   D. They have an expected lifetime of 20 to 30 years.

4. _____ Expanded polystyrene (EPS) foam is _____. (4.4-1)
   A. typically blue, pink, or green in color
   B. made up of small beads
   C. water-absorbent
   D. spun into fibers from molten rock

5. _____ *True or False?* Flashing tape can be used as a substitute for flashing throughout a building. (4.4-1)

6. _____ A window with one or more sashes that are hinged at the top and swing out at the bottom is called a(n) _____. (4.4-1)
   A. hopper
   B. fixed
   C. casement
   D. awning

7. _____ Which type of door is a popular choice for patio or backyard entries? (4.4-1)
   A. Door with sidelights
   B. Double door
   C. Door with transom
   D. Sliding door

8. _____ Most manufacturers of synthetic underlayment and housewrap now recommend the use of which type of fastener? (4.4-2)
   A. Plastic cap staples
   B. Roofing nails
   C. Galvanized staples
   D. Galvanized wood screws

9. _____ *True or False?* Roofing nails feature a larger head than framing nails. (4.4-2)

Lesson 4.5

1. _____ *True or False?* Aviation snips are designed to cut through sheet metal. (4.5-1)

2. _____ Aviation snips with _____ handles are designed for cutting curves to the left. (4.5-1)
   A. red
   B. yellow
   C. green
   D. blue

3. _____ A pneumatic or battery-operated tool that uses a coil of wire collated nails to attach shingles is a _____. (4.5-1)
   A. roofing nail gun
   B. cap stapler
   C. hammer tacker
   D. construction staple gun

Name ______________________________

Lesson 4.6

1. _____ *True or False?* Falls can occur from portable ladders on jobsites. (4.6-1)

2. _____ You should maintain _____ point(s) of contact when climbing in or out of machines. (4.6-1)
   A. no more than one
   B. at least two
   C. at least three
   D. four or more

3. _____ Fall protection is required anytime a worker has the potential to fall from a height of _____ or more. (4.6-2)
   A. 3′
   B. 4′
   C. 6′
   D. 8′

4. _____ What is an appropriate height for the top rail on a guardrail system? (4.6-2)
   A. 36″ from the working surface
   B. 42″ from the working surface
   C. 48″ from the working surface
   D. 52″ from the working surface

5. _____ A warning line system should be used _____. (4.6-2)
   A. whenever workers are working at a height of 6′ or more
   B. whenever workers are working at any elevation, regardless of height
   C. in conjunction with two or more conventional fall protection systems
   D. in situations where conventional fall protection systems are impractical

6. _____ *True or False?* Extension ladders are designed to be supported by leaning on a structure. (4.6-2)

7. _____ A guardrail system that protects an open side of a temporary stairway is a _____. (4.6-2)
   A. stair rail
   B. handrail
   C. step rail
   D. extension rail

Lesson 4.7

1. _____ Decimal fractions are fractions whose denominators are multiples of _____. (4.7-1)
   A. 2
   B. 5
   C. 10
   D. 100

2. _____ *True or False?* When you move one place to the right of the decimal point in a decimal fraction, the place value increases tenfold. (4.7-1)

3. _____ 9.82 + 16.5 + 132.24 = _____. (4.7-1)
   A. 157.111
   B. 157.560
   C. 158.560
   D. 158.010

4. _____ 52.6 ÷ 4.3 = _____. (4.7-1)
   A. 12.23
   B. 13.20
   C. 122.32
   D. 132.00

5. _____ *True or False?* 3/8 = 0.375 (4.7-2)

6. _____ *True or False?* When converting a mixed number to a decimal, the value of the whole number increases. (4.7-2)

7. _____ 0.35 = _____. (4.7-2)
   A. 3/50
   B. 7/20
   C. 7/50
   D. 35/10

Lesson 4.8

1. _____ *True or False?* Reference marks are used to help the builder orient the prints to one another. (4.8-1)

2. _____ *True or False?* Detail drawings are small-scale drawings. (4.8-1)

3. _____ *True or False?* Detail drawings may be shown either in plan view or elevation view. (4.8-1)

Lesson 4.9

1. _____ *True or False?* Water can infiltrate a structure through ice. (4.9-1)

2. _____ *True or False?* Concrete is impenetrable to water vapor. (4.9-1)

3. _____ To be classified as a vapor barrier, a material must have a permeability rating that is _____. (4.9-1)
   A. lower than 1
   B. lower than 5
   C. 5 or higher
   D. 10 or higher

4. _____ Vapor retarders have a perm rating that is _____. (4.9-1)
   A. 0
   B. between 0.1 and 1
   C. between 1 and 5
   D. 10 or higher

5. _____ *True or False?* High-performance buildings are less sensitive than buildings constructed decades ago. (4.9-2)

6. _____ A blower door test is used to _____. (4.9-3)
   A. test the strength of a door's locking mechanism
   B. test the integrity of a door's frame
   C. measure air movement through a building envelope
   D. measure variations in temperature in different rooms

7. _____ *True or False?* Thermal imaging cameras are useful for diagnosing HVAC system problems. (4.9-3)

Name ______________________________ Date ______________ Class __________

Activity 4-2: Skill Building—Practicing with Metalworking Tools

Learning Outcomes

After completing this lab activity, you will be able to:

- Use metalworking tools to prepare sheet metal for installation in a structure.

Introduction

During the dry in of a house, a builder will be working with different metal building materials such as flashing and aluminum cladding. These materials are needed to protect the home from air and moisture infiltration which can damage the structure of a home. The metal being used requires a different set of tools and skills from what a builder usually uses when working with wood. This activity will simulate typical operations needed to form and bend metals installed on a structure.

Text Reference

Chapter 4, *Drying In*

Equipment

The following tools and equipment are needed to perform these activities:

- Aluminum coil stock
- Aviation snips set: yellow, red, and green handles
- Carpenter's pencil
- Drafting compass
- Drill with a 3/8″ twist bit
- Framing square
- Safety glasses
- Sheet metal hand bender (seamer)
- Spring clamp
- Tin snips
- Utility knife
- Work gloves

Safety Review

Follow all safety procedures specified by your instructor and specified in any appropriate manufacturer's information. The following are some additional safety-related items to keep in mind:

- Wear safety glasses at all times.
- Be careful of sharp edges while handling metal pieces.
- Wear work gloves.
- Always retract the utility knife blade when not in use.

Procedure

1. Place the coil stock on a bench and roll out about 3′ to allow for room to work. Place the block of wood behind the roll or use the spring clamp on the edge of the roll to hold it in place.
2. Measure 10″ from the end of the roll and lay the framing square across the sheet at that mark. While holding the square tightly to the bench and square to the edge of the sheet, score the aluminum along the square with the utility knife. Apply some pressure but do not try and cut through the aluminum.

3. Remove the square from the roll and lift the metal off the bench at the score line. Bend the 10″ piece down toward the bench, creating a fold line along the score. Fold the piece all the way until it is touching the rest of the roll. Then fold it back up and over the score line. The aluminum should break off cleanly along the score line. If it does not break, fold it back and forth again until it does.

4. Roll the remaining coil stock back up and secure it so it does not unroll. Return it to its storage location or someplace where it is out of the workspace.

5. On the piece that was cut off, mark a line 8″ from one end and draw a square line across the 10″ width using the framing square. Cut this line using the tin snips, making sure to follow the line carefully.

6. On the remaining piece, mark a line again 8″ from the end and draw a square line across the piece. Cut this line carefully with the yellow-handled aviation snips. There should now be three pieces of aluminum that are 8″ × 10″.

7. On one of the 8″ × 10″ pieces, lay out a 5″ circle in the center of the piece. To find the center of the piece, lay the framing square from corner to corner diagonally and draw a line along the square. Move the square to the other corners diagonally and draw another line along the square. Those two lines intersect at the center of the piece that will locate the center of the circle. Use the compass to draw a 5″ circle.

8. Place the framing square across the circle at the center of the circle. Using the square as a guide, score a line with the utility knife inside the circle from one side to the other. Pick up the piece of aluminum and bend the piece back and forth at the score line until it breaks, creating an opening for the snips.

9. Using the yellow-handled aviation snips, cut out the circle drawn on the piece. Insert the lower jaw of the snips into the opening created by the cut and slide the snips to the end of the cut as shown in **Figure 1**. Start cutting from here and continue around the circle in a counterclockwise direction. Notice how the design of these snips works well with cutting in this direction.

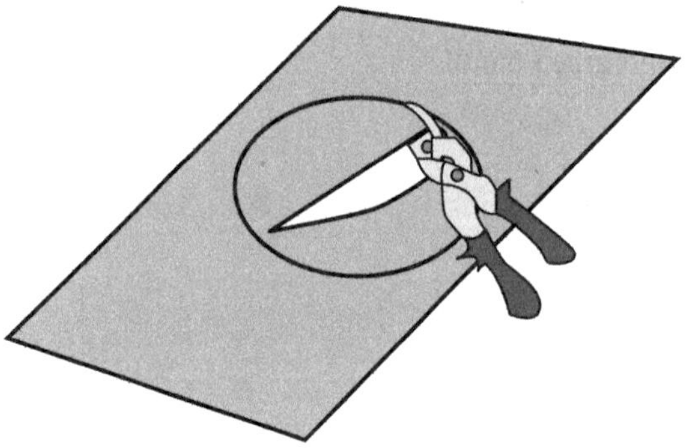

Goodheart-Willcox Publisher

Figure 1

10. Take the second 8″ × 10″ piece and draw a line down the center of the length to create two 4″ wide sides. From each end of the piece, draw a mark on the line at 2 3/4″. Using these two marks as center points, draw a 3″ circle at each location.

11. Cut a score line across one of the circles and break it open just like the first one. The red-handled aviation snips are designed to cut to the left. Insert the lower jaw in the cut and slide it to the end and start cutting the circle out in a counterclockwise direction. Notice how well these snips cut the circle compared to the yellow snips.

12. In the second circle on the same piece, drill a hole just inside the circle using the 3/8″ twist bit. Place the aluminum on a scrap piece of wood to support it while drilling. The green-handled snips are designed to cut to the right. Insert the point of the lower jaw of the snips into the hole as shown in **Figure 2** and cut out the circle in the clockwise direction. Start with short cuts until most of the cutting jaw fits into the opening being created. Try alternating hands when cutting to see which is more comfortable.

Name ______________________________

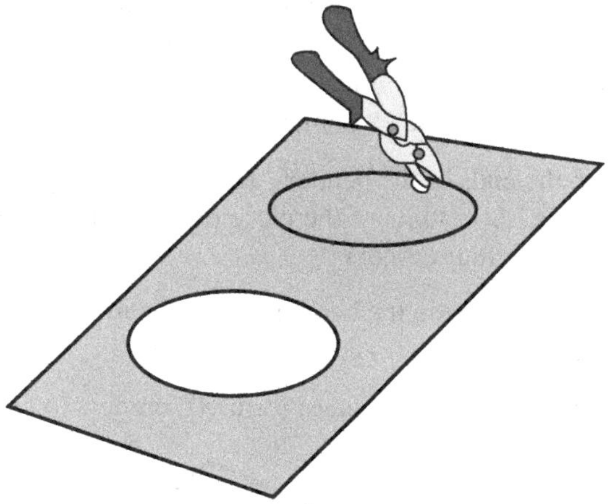

Goodheart-Willcox Publisher

Figure 2

13. On the last 8″ × 10″ piece, lay out a 6″ × 8″ rectangle centered on the piece as shown in **Figure 3**. Drill a 3/8″ hole inside each corner to start the cuts. Cut out the rectangle using the yellow-handled snips.

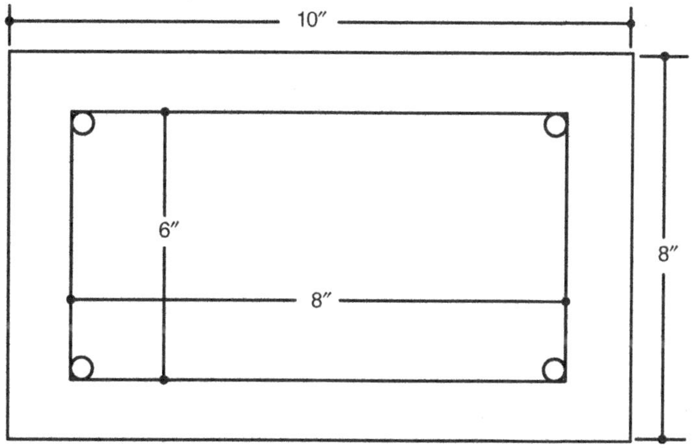

Goodheart-Willcox Publisher

Figure 3

14. On the 6″ × 8″ piece that was cut out, lay out lines on all four sides 1 1/4″ in from the edge as shown in **Figure 4**. At each end, cut out a small triangle at the intersection of the layout lines from the side as shown.

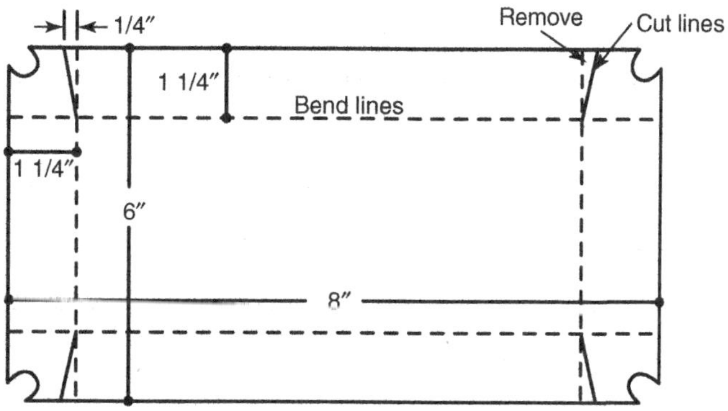

Goodheart-Willcox Publisher

Figure 4

15. Using the hand bender (seamer), bend each end up to form a 90° corner. Start on the end of the piece, line up the jaws of the bender with the line, and squeeze the handle tightly. Hold the aluminum piece on the bench while turning the bender up to form the angle. Bend it up and past 90° a little bit first and then back to the 90°. This will help the metal keep the desired angle as work continues.
16. On the end just turned up, line up the ends of the bender's jaws with the layout line to the triangle cut out. Squeeze tightly and bend it in toward the middle of the piece to form a tab at the corner. For now, bend it past 90° and leave it. Do this on each of the four corners.
17. Next, bend up the two sides of the piece using the bender. Again, bend the edge past 90° a little bit at first and then back to help form a tight corner.
18. Bend the tabs at each corner out to the sides by hand to form 90° corners to create a box as shown in **Figure 5**.

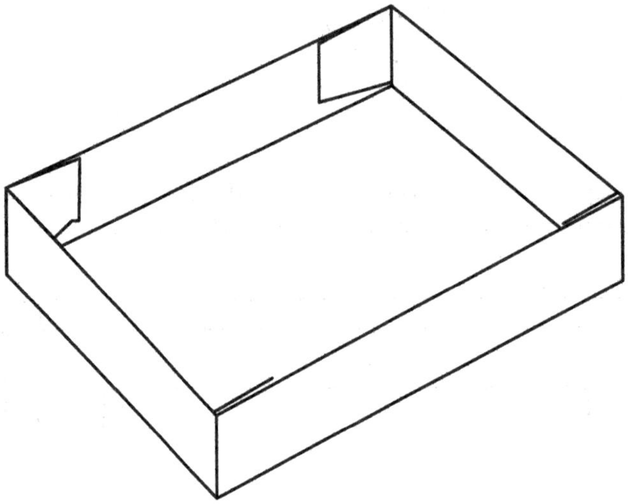

Goodheart-Willcox Publisher

Figure 5

Review and Analysis

1. Describe how the different-colored aviation snips are designed so they cut better in a specific direction. What about their shape helps accomplish this?

____

____

____

2. If a 45° bend were required on a piece of aluminum, what tool could be used to verify the correct angle was reached?

____

____

____

Name ______________________ Date ______________ Class __________

Activity 4-3: Skill Building—Waterproofing an Opening

Learning Outcomes

After completing this lab activity, you will be able to:

- Apply waterproofing materials to a window opening.

Introduction

The drying in stage of house construction is a critical part of protecting the house from water damage. One of the most susceptible locations where water damage occurs is around windows and doors. If not sealed properly, water gets past the siding and trim and finds its way into the wood framing.

In this activity, you will practice installing membrane flashing tape in an opening to prevent water penetration. The first rule of flashing is that it is always installed from the bottom up. This ensures water always runs down on top of the next layer and eventually out of the structure.

Text Reference

Chapter 4, *Drying In*

Equipment

The following tools and equipment are needed to perform these activities:

- Carpenter's pencil
- J-roller
- Membrane flashing tape, 6″ or 9″ wide (6″ if using a 2×4 wall, 9″ if 2×6)
- Membrane flashing tape, flexible, 6″ or 9″ wide (6″ if using a 2×4 wall, 9″ if 2×6)
- Safety glasses
- Staple gun or hammer tacker
- Utility knife
- Wood bevel siding

Safety Review

Follow all safety procedures specified by your instructor and specified in any appropriate manufacturer's information. The following are some additional safety-related items to keep in mind:

- Wear safety glasses at all times.
- Retract the blade of a utility knife when not in use to prevent punctures or cuts.

Procedure

This activity requires a wall framed with a window opening and the weather-resistive barrier (WRB) installed. The WRB should cover the wall and span over the opening.

1. Cut the WRB in the window opening to prepare for the flashing. Cut completely across the top and the bottom at the edge of the opening, and then cut in the center of the opening from top to bottom so the cuts look like the letter *I*.
2. In the center of the window opening, bend each piece of WRB from the center into the opening, and then cut it off down the jack stud 4″ in from the face of the wall. Do this for both sides. Staple the flap against the jack stud, ensuring the staples are set flat to the wall.
3. At the top of the opening, cut the WRB at a 45° angle up 4″ away from each corner. Do not remove this. This flap will go over the nailing flange on a window or over the cap flashing on a door after installation to ensure water runs out and away from the wall. Refer to **Figure 1**.

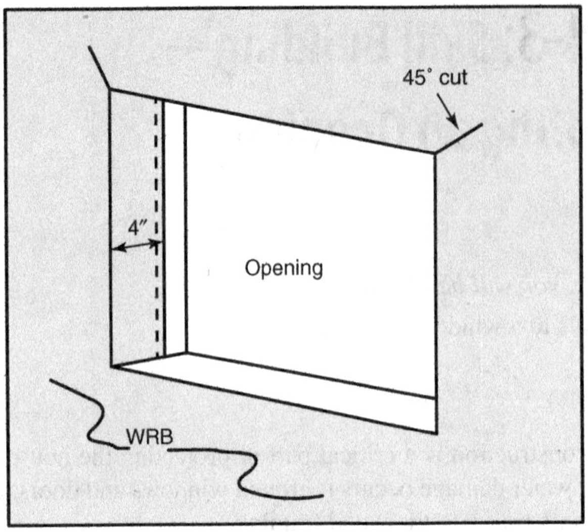

Goodheart-Willcox Publisher

Figure 1

4. Install wood bevel siding at the bottom of the opening to create a sloped surface toward the outside. If water enters the space around the window, it will run down to the bottom of the opening and to the outside of the wall. Cut the piece to be the same length as the window opening. Nail it in place with the thin edge of the siding flush with the front edge of the wall.
5. The first piece of flashing tape will go on top of the wood siding piece at the bottom of the opening and at least 6″ up each side of the opening. Measure the width of the opening and add 6″ for each side; then, use the utility knife to cut this total length from the roll of flashing tape.
6. Install the first piece of flashing tape to be flush with the inside edge of the wall and fold the tape extending past the wall onto the surface of the WRB to seal the opening. First, peel the paper backer off the flashing tape to expose the wider part of the sticky side. Bend 6″ on one end back onto itself exposing the sticky side. Place the bend into the corner of the opening on one side and press it tightly into the corner. Keep the other end up, away from the sill so it does not stick to anything. Line up the other end with the inside of the wall and slowly press it into place.
7. Slide your hand from corner to corner to ensure a smooth surface. Press the flashing tape tight into the corners and then press the side pieces up against the walls of the opening.
8. At both corners of the opening, cut the flashing tape along the side wall to create a tab at the end. Lay the tab out of the opening on the front of the wall. Use **Figure 2** as a guide.

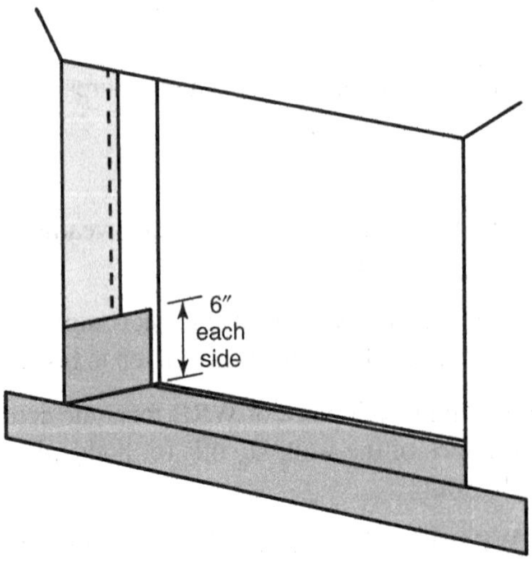

Goodheart-Willcox Publisher

Figure 2

9. Remove the paper backer from the horizontal piece hanging out of the opening and bend it over the WRB without putting ripples in the WRB. Start bending it over at the center of the opening and slowly work toward each corner.

10. When all surfaces are bent over to the front of the wall, roll all the flashing tape with the J-roller to ensure a good bond.

Installing Flashing Tape at Corners for Full Protection

At the bottom corner of the opening where the flashing tape has been cut, there may be a small hole left at the intersection that needs to be sealed. There are two possible ways of installing flashing tape in corners to accomplish complete protection. The next steps will give you the opportunity to practice each method.

Method 1—Bottom-Left Corner

1. Cut a 6″ piece of the flashing tape and trim it in an hourglass-shaped piece that narrows to 1″ at the center, **Figure 3**.

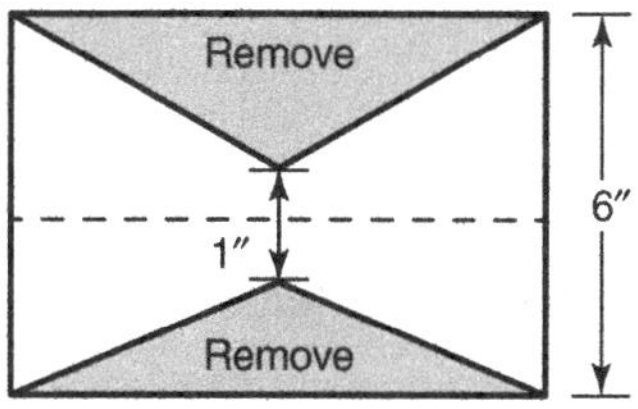

Goodheart-Willcox Publisher

Figure 3

2. Fold the piece along the dotted line to create a crease and then remove the paper backer.

3. Place the piece in the bottom-left corner of the opening and line up the narrow center of the hourglass shape on the edge of the wall.

4. Press the piece into the corner and on either side of the opening corner. Press a finger tight into the outside corner at the face of the wall, and with the other hand pull the end out away from the corner and fold it back against the wall, **Figure 4**. Flatten it out by hand as best as possible and then use the J-roller.

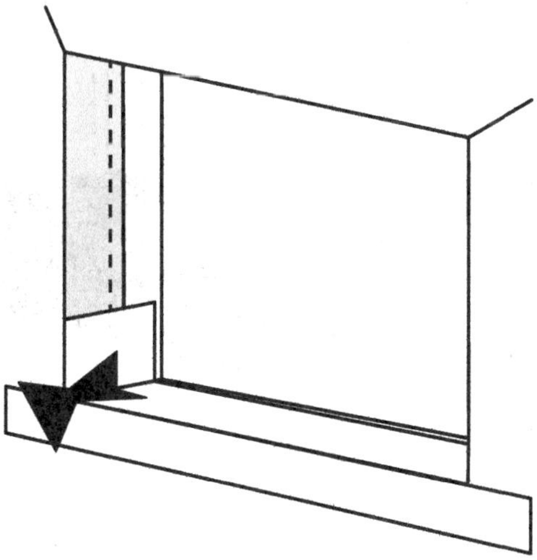

Goodheart-Willcox Publisher

Figure 4

5. Cut another piece of flashing to be 10″ long. Fold it over at 6″ from one end and remove the paper backer.

6. Press the bend into the corner on top of the hourglass piece to be even with the inside face of the wall. Cut the flashing along the bottom of the opening, starting at the corner and out to the right. Bend the flashing out and against the wall to the side of the opening overlapping the end of the first piece to look like **Figure 5**.

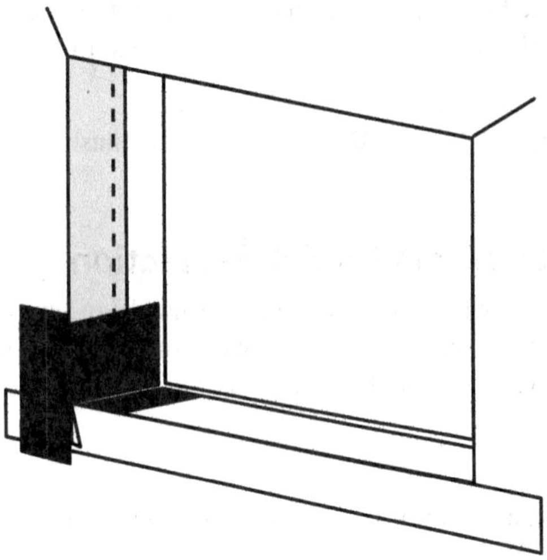

Goodheart-Willcox Publisher

Figure 5

Method 2—Bottom-Right Corner

1. Cut a 12″ piece of flexible flashing tape from the roll. Remove the paper backing and fold it in half with the sticky side facing out.
2. Place the fold into the right bottom corner of the opening and press it tightly into the corner. Then open the piece up and press it tight to the bottom and side of the opening.
3. Starting at the inside of the corner, push the flexible flashing tape out and down over the edge of the opening. Rub it a bit in each direction, causing it to expand as it reaches the surface of the wall. Keep rubbing it out and away from the corner until it is flat against the wall and adhered to the WRB. Roll it for a good bond. **Figure 6** shows the finished job.

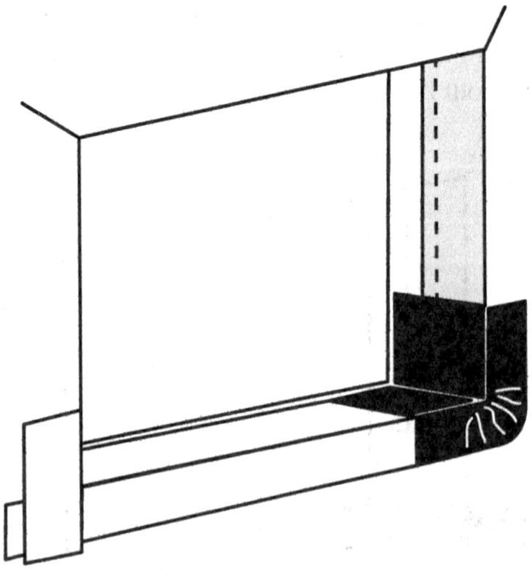

Goodheart-Willcox Publisher

Figure 6

4. This completes the flashing required to prepare the bottom of an opening whether it is a door or a window. The sides and top will be flashed at a later stage of installation.

Name ______________________________

Review and Analysis

1. What is the purpose of the wood bevel siding installed at the bottom of the opening?

____

____

____

2. Why was the hourglass-shaped piece of flashing placed at the corner?

____

____

____

3. How will the flap created in the WRB by the 45° cuts at the top corners be used later?

____

____

____

Notes

Name ______________________ Date ____________ Class __________

Activity 4-4: Skill Building—Flashing a Roof

Learning Outcomes

After completing this lab activity, you will be able to:

- Install step flashing at a roof and wall intersection.

Introduction

The drying in process ensures the house is watertight and protected from the weather. One of the steps for drying in a home is the installation of step flashing where a roof intersects a wall. This location is susceptible to water damage because rainwater or snow will collect here and may seep in and cause wood damage.

When step flashing is installed, it provides a path to allow the water to find its way down and off the roof without seeping into the wall materials. This exercise will show how step flashing is installed to accomplish this.

Instructor Note: A small section of a sloped roof is required with a vertical wall at one side of the roof. This can be built close to the floor so no fall protection would be required. The roof can be prepared ahead of time with the ice and water barrier and drip edge installed ready for shingles.

Text Reference

Chapter 4, *Drying In*

Equipment

The following tools and equipment are needed to perform these activities:

- Aluminum flashing
- Aviation snips
- Drip edge
- Gloves
- Hammer
- Roofing nails
- Safety glasses
- Shingles
- Stapler
- Starter strip shingles
- Utility knife

Safety Review

Follow all safety procedures specified by your instructor and specified in any appropriate manufacturer's information. The following are some additional safety-related items to keep in mind:

- Wear safety glasses at all times.
- Wear work gloves when handling sheet metal to protect from sharp edges.

Procedure

Any type of flashing is installed with the same goal in mind: it directs the water down on top of another waterproof surface until it reaches the ground therefore keeping the structure dry. Roofing materials and flashings are always installed from the bottom of the roof up to the top to accomplish this.

1. The first materials installed on a roof surface are the drip edge and an ice-and-water barrier or an underlayment. This first layer provides an additional protective layer against the weather. During construction, it provides protection before the finished roofing materials can be installed. Staple the underlayment down with 5″ folded against and going up the wall surface, **Figure 1**.

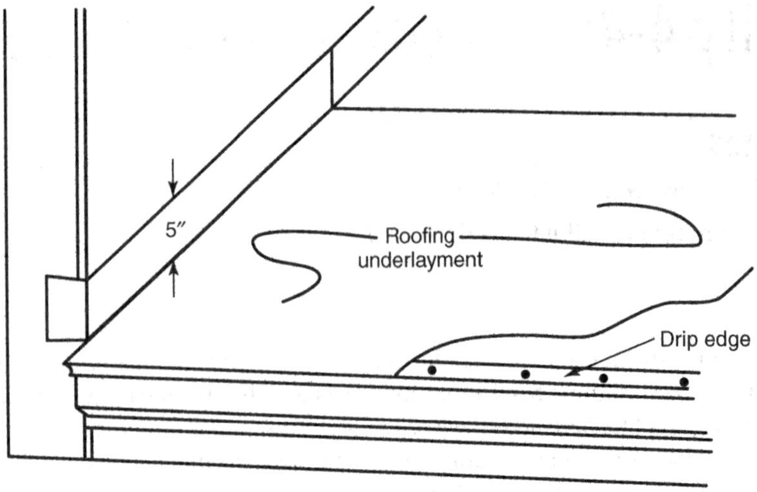

Goodheart-Willcox Publisher

Figure 1

2. Next, install a shingle starter strip. The starter strip provides an edge with a tar strip that will adhere to the first full shingle row installed. This keeps the wind from picking up the edge of the shingle and causing damage to the roofing.
3. Then, install the first piece of step flashing. Cut a piece of aluminum flashing 10″ × 8″ using snips or a utility knife. Bend the piece in half to produce an *L* shape with 5″ of metal on each side. Bend it so the angle is a little larger than 90° so it will be tight against the wall when pushed in and fastened.
4. At one end, cut along the bend for 3″ with aviation snips. Place the piece at the bottom of the roof against the wall with the end of the cut at the corner. Bend the piece that goes past the corner back to the end of the wall. To make a nice precise bend, take the piece and bend the tab like folding a piece of paper tight to the other side. Then, open the tab back out to make a 90° for the wall corner.

> **NOTE**
> Out on the jobsite, a small amount of roofing tar would be placed at the corner intersection to seal it well. To save material and avoid a mess, do not apply the roofing tar in the lab.

5. Hold the flashing in position at the corner, pushing it tight against the wall and tight up against the corner. Hammer two nails into the roof, about 3″ from the wall, to hold the flashing in place, **Figure 2**. Do not put any nails in the wall.

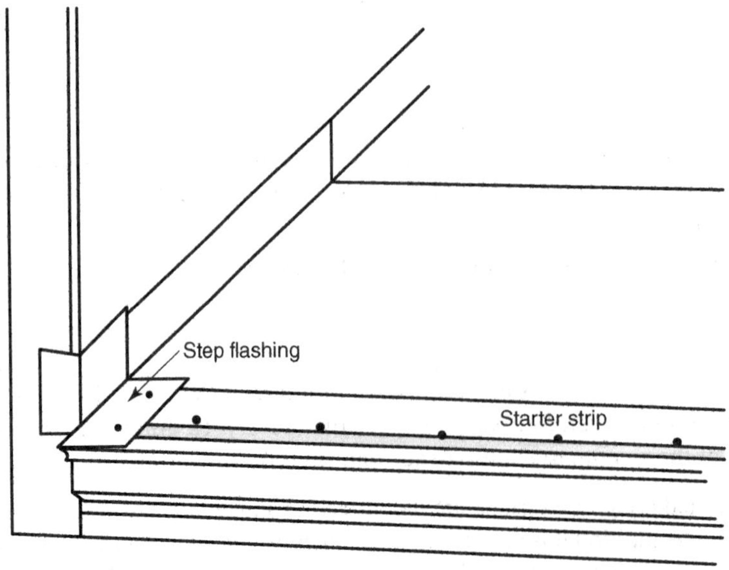

Goodheart-Willcox Publisher

Figure 2

Name ____

6. Place the first full shingle next. Do not push the end of the shingle tight into the corner of the first step flashing. Start the shingle about 1/4″ from the wall to allow for a channel to be created for easy flow of rainwater down and off the roof. Nail the shingle in place as instructed by the manufacturer.

7. Cut a second piece of aluminum step flashing and bend it like the first piece in Step 4. Place the flashing on the roof against the wall. Place just above the exposure height of the shingle so it will not be visible when the next shingle goes on top of it. Push the flashing tightly against the wall and hammer two nails in the roof side again, **Figure 3**.

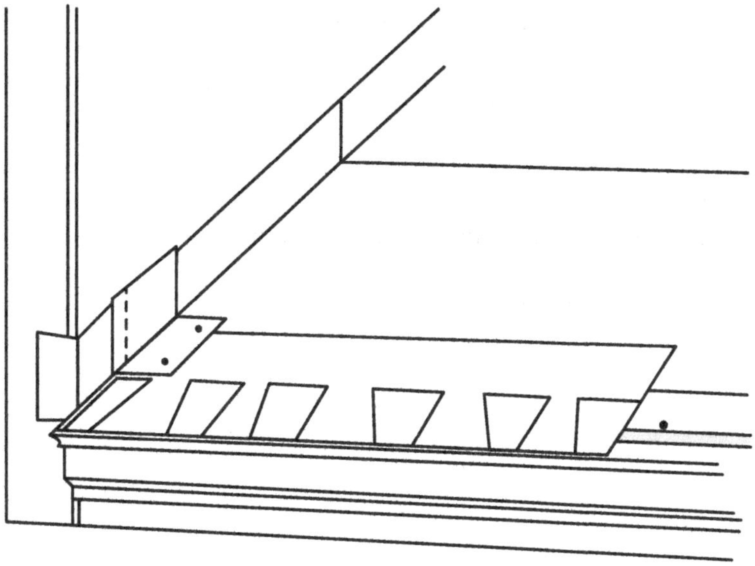

Goodheart-Willcox Publisher

Figure 3

8. This process is repeated as each row of shingles is installed, **Figure 4**. Be sure to push the flashing tight into the corner as it is nailed to ensure there is no opening along the wall for water to go into during a storm.

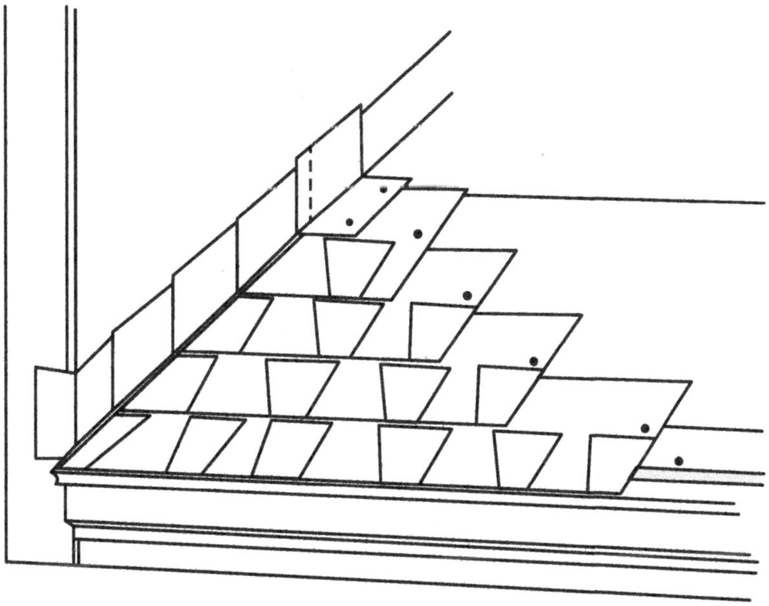

Goodheart-Willcox Publisher

Figure 4

Review and Analysis

1. What is the purpose of a starter strip along the bottom edge of the roof?

____

____

____

2. Describe the nailing pattern for the shingle from the manufacturer.

____

____

____

3. Why is the step flashing installed at or above the exposure line of the next shingle?

____

____

____

Name ______________________ Date ____________ Class ____________

CHAPTER 5 Exterior Finishes

Activity 5-1: Chapter 5 Review

Answer the following questions after studying the chapter.

Lesson 5.1

1. _____ *True or False?* If you are looking for opportunities for growth in your career, you will often have to seek them out on your own. (5.1-1)

2. _____ *True or False?* Continuing education and research provide an avenue for growth into other trade specialties. (5.1-2)

3. _____ *True or False?* You must join a formal networking group or other professional organization in order to engage in networking. (5.1-2)

4. _____ Which statement about goal setting is true? (5.1-3)
   A. If you miss the mark on a goal, it means you are not cut out for a career in residential construction.
   B. It is important to regularly reflect on your progress toward your goals.
   C. The goals you set for yourself should always be long-term goals.
   D. Goal-setting becomes less important as you become a more experienced worker.

Lesson 5.2

1. _____ Which type of law is specific to a geographic area within a municipality? (5.2-1)
   A. Environmental regulation
   B. Labor law
   C. Tax law
   D. Zoning ordinance

2. _____ A law that controls the height of new construction is an example of a(n) _____. (5.2-1)
   A. consumer protection law
   B. environmental regulation
   C. labor law
   D. zoning ordinance

3. _____ Which type of law might require contractors to provide written contracts? (5.2-1)
   A. Consumer protection law
   B. Labor law
   C. Tax law
   D. Zoning ordinance

4. _____ OSHA standards are an example of _____. (5.2-1)
   A. consumer protection laws
   B. environmental regulations
   C. labor laws
   D. licensing requirements

5. _____ Taxes may be levied against a contracting business at which levels? (5.2-1)
   A. Local and state
   B. State and federal
   C. Local, state, and federal
   D. Local, state, federal, and international

Lesson 5.3

1. _____ Which feature of a home is part of the exterior finish? (5.3-1)
   A. Closet
   B. Garage
   C. Kitchen
   D. Porch

2. _____ *True or False?* During the construction process, the siding is generally applied before the cornice of the home is finished. (5.3-2)

3. _____ *True or False?* The fascia is horizontally attached along the ends of the rafters. (5.3-2)

4. _____ Which component closes off the space between the wall and the fascia board? (5.3-2)
   A. Soffit
   B. Frieze board
   C. Cladding
   D. gambrel

5. _____ A decorative pipe or tube through which water from the roof is directed toward the ground is a _____. (5.3-2)
   A. cornice
   B. downspout
   C. fascia
   D. rainscreen

6. _____ Siding is an example of a(n) _____ system. (5.3-2)
   A. decking
   B. insulation
   C. cladding
   D. cornice

Lesson 5.4

1. _____ Aluminum coil stock is commonly used for which purpose? (5.4-1)
   A. Roofing
   B. Fascia trim
   C. Insulation
   D. Gutter systems

2. _____ PVC trim boards are _____. (5.4-1)
   A. difficult and expensive to maintain
   B. susceptible to moisture and rot
   C. an alternative to hardwood flooring
   D. appropriate for use as exterior trim

3. _____ *True or False?* Fiber cement siding is subject to damage by insects. (5.4-1)

4. _____ *True or False?* Exterior insulation and finish systems offer high R-values. (5.4-1)

5. _____ Cultured stone is adhered to a wall using _____. (5.4-1)
   A. mortar
   B. masonry nails
   C. masonry screws
   D. industrial adhesive

6. _____ *True or False?* Pressure-treated lumber expands as it dries. (5.4-1)

7. _____ Which statement about composite decking is true? (5.4-1)
   A. It is made entirely of recycled plastics.
   B. It is prone to warping and fading.
   C. It must be sanded and stained after installation.
   D. It is durable and low-maintenance.

8. _____ Which statement about deck screws is true? (5.4-2)
   A. You must drill a pilot hole before installing them.
   B. They feature a rounded head.
   C. Many types are coated with ceramic, polymer, or epoxy.
   D. They have either ring or spiral shanks.

Lesson 5.5

1. _____ What is a seamer used for? (5.5-1)
   A. Bending thin metal flashing
   B. Removing vinyl siding
   C. Forming locking tabs on vinyl siding
   D. Cutting irregular shapes on edges

Name ______________________________

2. _____ *True or False?* A vinyl siding removing tool has a hooked blade. (5.5-1)

3. _____ *True or False?* A router uses a collet-type chuck to hold a bit. (5.5-1)

4. _____ Straight bits are used on a router for _____. (5.5-1)
   A. creating round corners
   B. forming beveled edges
   C. cutting dados
   D. trimming edges

5. _____ *True or False?* When using a router to make a cut along an edge, you should feed it from right to left. (5.5-1)

Lesson 5.6

1. _____ The midrail on scaffolding should be located _____. (5.6-1)
   A. exactly at the worker's waist height
   B. at least 45″ above the working surface
   C. no more than 18″ above the working surface
   D. half-way between the working surface and top rail

2. _____ A supported scaffold must be stabilized if its height-to-base ratio is greater than _____. (5.6-1)
   A. 2:1
   B. 4:1
   C. 5:1
   D. 8:1

3. _____ A mudsill is used when _____. (5.6-1)
   A. roof-mounted scaffolding is used
   B. the legs of scaffolding are set on solid concrete
   C. the legs of scaffolding are set on a compactable surface
   D. the legs of scaffolding are different lengths

4. _____ *True or False?* Metal scaffolding platforms can extend more than 8′ between supports. (5.6-1)

5. _____ *True or False?* Ladder jacks are specifically designed to support low platforms for interior work. (5.6-1)

6. _____ *True or False?* The anchor points on an aerial lift meet the impact requirements for a personal fall arrest system. (5.6-1)

Lesson 5.7

1. _____ *True or False?* When something is plumb, it is perfectly horizontal. (5.7-1)

2. _____ *True or False?* The relative distance from the bottom or top of a window should be the same for all like-sized windows. (5.7-1)

3. _____ *True or False?* It is easier to level over long spans with a laser level than with a spirit level. (5.7-1)

4. _____ A triangle that has a90° angle is called a(n) _____ triangle. (5.7-2)
   A. equilateral
   B. isosceles
   C. obtuse
   D. right

5. _____ If side *a* of a right triangle is 8′-6″ long and side *b* is 10′-9″ long, how long is the hypotenuse? (5.7-2)
   A. 13′-5.0″
   B. 13′-6.8″
   C. 13′-8.4″
   D. 13′-8.8″

6. _____ When using the simplified method to verify that a corner is 90°, you should measure out the two sides of the right triangle to what distances? (5.7-2)
   A. 3′ and 5′
   B. 4′ and 6′
   C. 5′ and 7′
   D. 6′ and 8′

Lesson 5.8

1. _____ *True or False?* The various elevations on elevation drawings are generally oriented to the site by labeling them according to the direction they face. (5.8-1)

2. _____ By studying the elevation views, a builder can determine _____. (5.8-1)
   A. roof slopes
   B. bedroom wall colors
   C. interior vent locations
   D. bathroom fixture locations

3. _____ What does this symbol represent in construction drawings? (5.8-2)

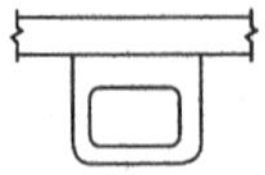

   A. Built-in lavatory
   B. Dental lavatory
   C. Kitchen sink
   D. Wall lavatory

4. _____ *True or False?* A list of abbreviations used on a set of construction drawings is included on every page of the drawings. (5.8-2)

5. _____ What does the abbreviation DR commonly stand for on construction drawings? (5.8-2)
   A. Dining room
   B. Door
   C. Drain
   D. Driveway

Lesson 5.9

1. _____ *True or False?* A home that does not leak air is far more energy efficient than one that does. (5.9-1)

2. _____ *True or False?* One goal of reducing energy consumption is to increase thermal bridging. (5.9-1)

3. _____ *True or False?* Low-impact materials are produced distant from the jobsite and are then trucked in when needed. (5.9-2)

4. _____ When VOCs enter the air, they react with other elements to produce _____. (5.9-2)
   A. carbon dioxide
   B. nitrous oxide
   C. ozone
   D. methane

5. _____ *True or False?* Precast concrete foundation walls generate little to no waste compared to concrete walls that are cast on site. (5.9-2)

Name ______________________________ Date ______________ Class ____________

Activity 5-2: Skill Building—Bending Materials with a Brake

Learning Outcomes

After completing this lab activity, you will be able to:

- Use the portable field brake to prepare aluminum cladding.
- Install aluminum cladding to cover exterior trim.

Introduction

When a siding installer completes the exterior finishes of a house there are many areas where exposed framing lumber or wood trim need to be protected from the weather or given a finished appearance to match the rest of the house. Aluminum coil stock is used to accomplish this protection. Coil stock is available in many colors and comes in a large roll of flat aluminum.

The installer cuts and bends the metal to fit around the parts of the house that need protecting. A portable field brake is the tool an installer must become proficient in using to create pieces that cover the trim. This activity will start developing basic skills for this process.

Text Reference

Chapter 5, *Exterior Finishes*

Equipment

The following tools and equipment are needed to perform these activities:

- 1 1/2″ × 1 1/2″ × 44″ lumber
- Aluminum coil stock
- Aviation snips
- Carpenter's pencil
- Colored trim nails
- Framing square
- Hammer
- Hand bender
- Portable brake
- Safety glasses
- Speed square or combination square
- Utility knife

Safety Review

Follow all safety procedures specified by your instructor and specified in any appropriate manufacturer's information. The following are some additional safety-related items to keep in mind:

- Wear safety glasses at all times.

Procedure

1. Prepare a frame using the 1 1/2″ × 1 1/2″ lumber on a wall surface at a comfortable height to work. Cut and fasten the lumber on the wall according to **Figure 1** as a mock window frame to wrap with aluminum. Be sure the frame is level and square as it is fastened to the wall.

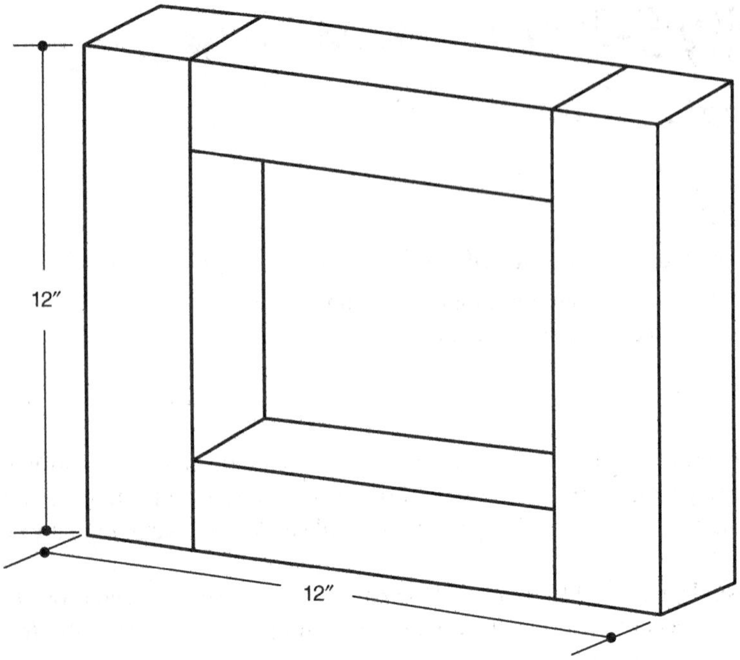

Goodheart-Willcox Publisher

Figure 1

2. Cut a 12″ piece of coil stock from the roll using the method practiced in Activity 4-2.
3. Install finish cladding pieces from the bottom up, just as was done with the flashing, so water will always run off and away from the building. Carefully measure the thickness of the bottom piece from the wall and the width of the face of it. Add the width of all three faces together to establish how wide to cut the first piece of metal. It should be 4 1/2″.
4. On one end of the piece of aluminum, measure 4 1/2″ along the edge and place a small mark with the pencil. Repeat the measurement and mark on the other edge.
5. Open the jaws of the brake, insert the aluminum piece into the jaws, and position the piece so the two marks made in Step 4 line up with the upper jaw. Lightly clamp the jaws down and double check that the marks are still lined up with the edge of the jaw. Adjust the piece as needed to align the marks. Once the piece is lined up accurately, clamp the jaw down to its full position.
6. Use the utility knife to cut this piece off. Put the knife tight against the upper jaw, into the piece of aluminum, and make a score cut. Stop the cut before coming to the other end of the piece. Never run the knife all the way off the other end and into the jaws of the brake because this will ruin the brake hinge at the jaws. Cut close to the end, then remove the knife and turn it around to make a score cut from that end back to meet the first cut.
7. Grab the bending handles and lift the bending hinge up almost to the top. Let the bending handles back down. Bend the piece of aluminum back down by hand and the piece should break off. If it does not break off, repeat bending the piece up and back down by hand. This should produce a piece of aluminum 4 1/2″ × 12″. Open the jaws of the brake, remove the rest of the aluminum piece, and set those pieces aside for later.
8. This piece will be bent into a U shape that will fit over the bottom of the frame. On the back of the piece or the inside of the bends, mark 1 1/2″ from the edge at both ends. This can be done from both edges now to save some marking time later. Refer to **Figure 2**.

Name ____

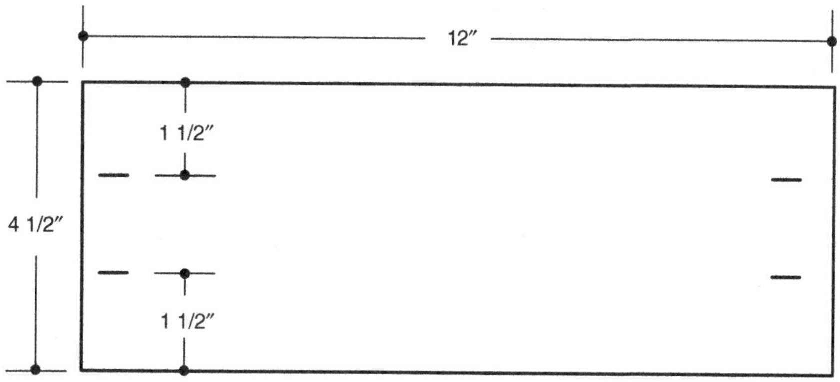

Goodheart-Willcox Publisher

Figure 2

9. Put the piece into the jaws of the brake with about the 1 1/2″ out past the jaws and lightly close the jaws. Adjust the piece to line up the 1 1/2″ marks with the edge of the upper jaw and then clamp the jaws fully closed. Pull the bending handles up until the aluminum is at a 90° angle to the face.

10. Open the jaws and turn the piece around to the other set of marks. Align them to the jaw edge and clamp in place. Bend this side up to the 90° angle to be the same. Unclamp the jaws and remove the piece. This should be a nice square U-shaped channel.

11. Line up the piece against the bottom of the frame with the ends and the top of the wood, **Figure 3**. Mark the inside edges of each side of the frame. Using the square, lightly draw a square line across the top face of the piece.

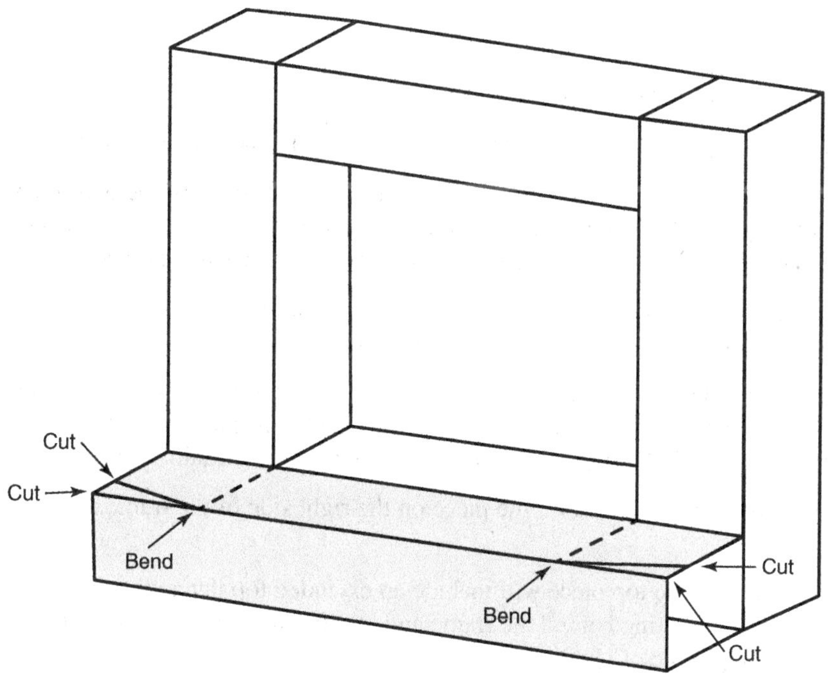

Goodheart-Willcox Publisher

Figure 3

12. Cut along the corner bend to the square mark on each end as shown in **Figure 3**. Then, cut out a small triangle along the front edge of the tab as shown. This will make sure the corner of the tab will not interfere with the side pieces when installed.

13. Use the hand bender to bend the tab created up to a 90° corner on each side. This tab acts as flashing inside the corner protecting the wood underneath and also ensures the color of the aluminum trim will be seen inside the corner when completed.

14. Slide the aluminum piece over the bottom wood frame piece. Using a colored trim nail, fasten the aluminum to the frame on the bottom of the frame, **Figure 4**. Nailing here puts the nail in a place that would not be seen when the siding is completed.

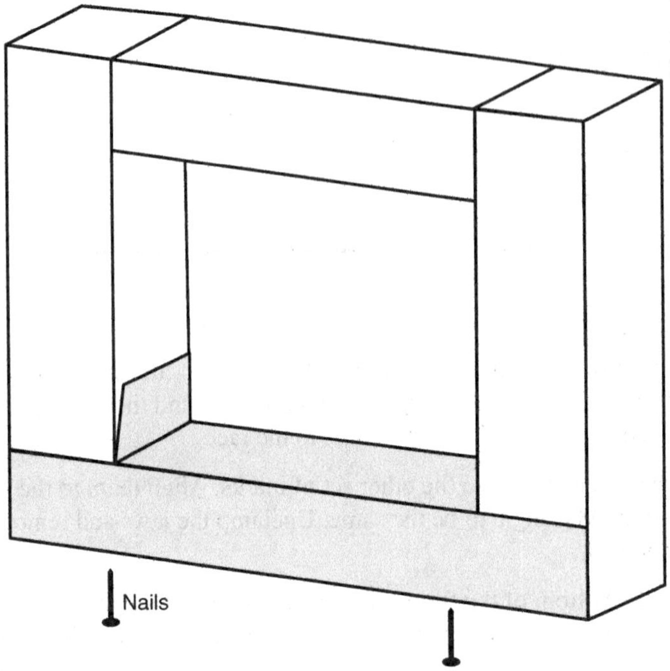

Goodheart-Willcox Publisher

Figure 4

15. Repeat Steps 2–7 to measure and cut two new pieces.
16. Repeat Steps 8–10 two more times to make two more U-shaped channels that will cover both sides of the frame.
17. Repeat Step 11 but hold one piece against the left side of the frame instead of the bottom. Mark the two corners again. The top of the piece will be cut to create a tab just like with the bottom. Bend the tab to a 90° corner with the hand bender to ensure the finished color is seen inside the corner when complete.
18. On the bottom corner, using the square, mark a 45° line across the piece from the mark to create what will look like a mitered corner when installed. Mark a square line across the inside face of the piece from the corner mark.
19. Start cutting at the inside edge of the square line and cut up to the corner mark. Then, bend that tab up to allow the snips to turn and cut across the front face at the 45° line. Slide this piece on to the frame covering the tab at the bottom corner, **Figure 5**. Nail this piece on the outside of the frame again so the nails will be covered later.
20. Repeat the same marking steps and cuts for the piece on the right side of the frame and fasten on the outside of the frame again.
21. To finish covering the frame, the top piece will include an extended top flange that provides flashing for the frame and keeps water from getting behind the frame and into the wall. The flange will extend 3″ above the frame. From the remaining piece of aluminum, cut a piece measuring 7 1/2″ × 12″.
22. On the back of the aluminum piece, put two marks that are 1 1/2″ from one edge as done in Step 8. From the same edge, put two marks at 3″ to mark the second bend. Bend these two corners in the brake to create a *J* shape.
23. On the front or colored side of the *J*, measure in 1 1/2″ from the second bend and mark both ends for the final bend. This final bend will create the flashing part of the aluminum piece that will go up the wall.
24. Place the piece in the brake lining up the marks with the upper jaw and clamp it down. Lift the bending handles to form a 90° corner. The piece should look like **Figure 6**.

Name ____

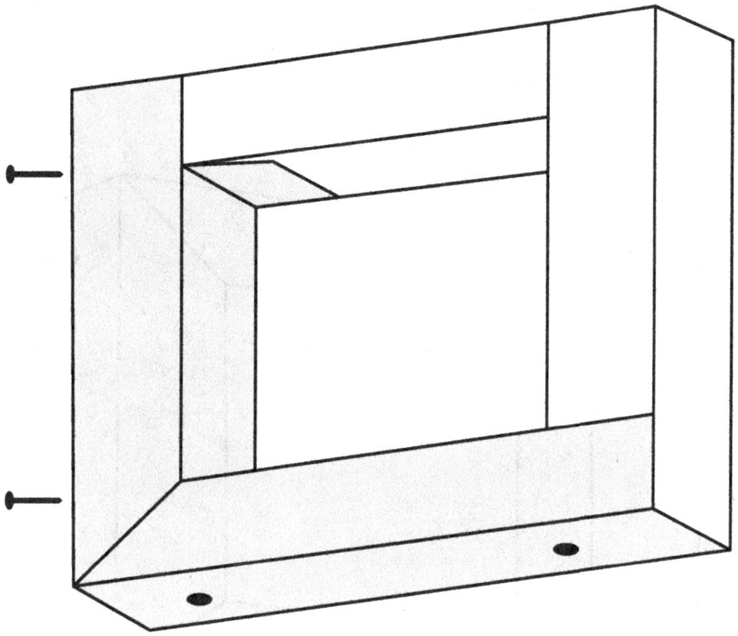

Goodheart-Willcox Publisher

Figure 5

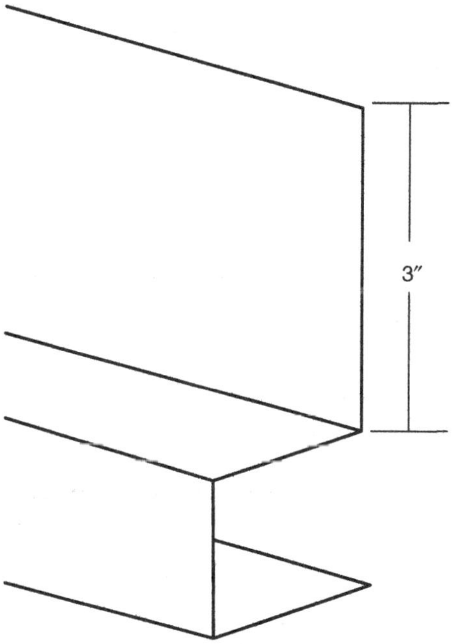

Goodheart-Willcox Publisher

Figure 6

25. Hold this piece against the top of the frame and mark the inside corners. This piece will have both corners cut at a 45° angle to look like mitered corners. Slide the piece over the tabs in the corners and place it tight against the wall. This piece should be nailed to the wall above the frame so no holes are created in the top of the frame where water may get in.

26. Nail the bottom of this piece inside the frame to keep it tight. Drive a nail up into the inside of the frame to secure it. Be careful nailing here so you do not dent the aluminum. The finished frame should look like **Figure 7**.

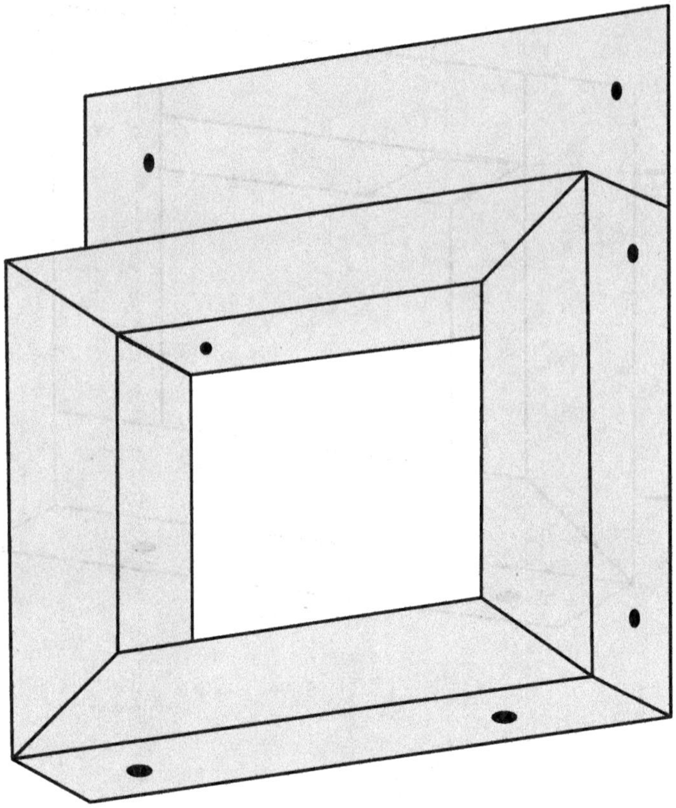

Goodheart-Willcox Publisher

Figure 7

Review and Analysis

1. What was the purpose of nailing the cladding pieces on the outside of the frame?

2. It is very easy to put dents in the aluminum as the nails are placed. What could be done to prevent creating dents while nailing?

3. Why should nails not be placed on the top of the frame when fastening the cladding?

Name ______________________ Date ____________ Class __________

Activity 5-3: Skill Building— Working with Siding Tools

Learning Outcomes

After completing this lab activity, you will be able to:

- Measure and cut vinyl siding and accessories.
- Install vinyl siding using the accessories as designed.

Introduction

Vinyl siding is the most common exterior finish used in residential construction. There are contractors across the country who specialize in and only install vinyl siding. An advantage of this siding is that it is easy to maintain after installation.

Vinyl expands and contracts with changes in the outside temperature which must be anticipated when installing all accessories and siding panels. In this activity, installation steps will be practiced with attention to how allowances are made to anticipate this movement.

Instructor Note: Use the frame wrapped with aluminum from Activity 5-2 this activity. It will be like a window being worked around.

Text Reference

Chapter 5, *Exterior Finishes*

Equipment

The following tools and equipment are needed to perform these activities:

- Aviation snips
- Carpenter's pencil
- Hammer
- J-channel
- Level, 2′
- Safety glasses
- Siding removal tool or zip lock tool
- Snap lock punch
- Square
- Tin snips
- Utility trim/undersill trim
- Vinyl siding, double 4 panel

Safety Review

Follow all safety procedures specified by your instructor and specified in any appropriate manufacturer's information. The following are some additional safety-related items to keep in mind:

- Wear safety glasses at all times.
- Be careful of sharp edges on the siding and accessories while cutting.

Procedure

Since vinyl siding panels and accessories expand and contract during changes in temperature, the first thing to understand is how to place the nails used. The siding must be able to move. The nails are never set tight against the vinyl and the wall. There should be about 1/32″ space between the head of the nail and the vinyl nailing flange, **Figure 1**. If the nail is too tight, the siding will buckle when it expands.

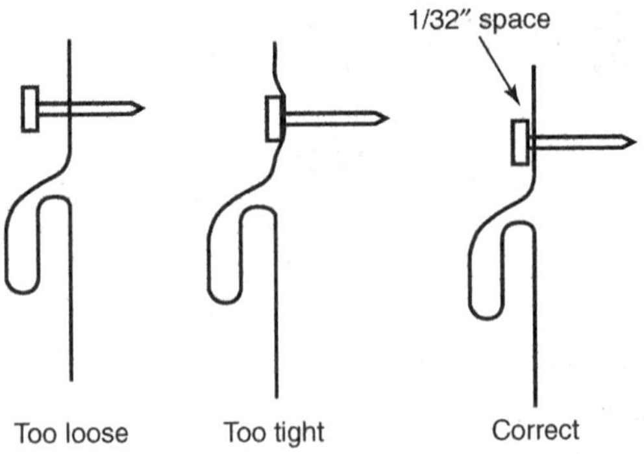

Goodheart-Willcox Publisher

Figure 1

All siding panels and accessories have a nailing slot along the edge for fastening. The nail should be placed in the center of that slot to allow for movement, **Figure 2**. If the nail is not centered, it may limit the movement of the siding and may also cause the panel to buckle on the wall.

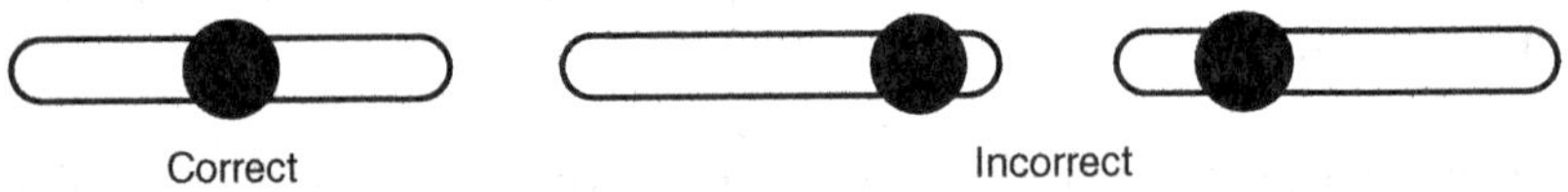

Goodheart-Willcox Publisher

Figure 2

1. The first step in the siding installation process is to prepare the windows and doors with J-channel around the outside of the frames. This J-channel will hide the ends of the siding which will be shorter than the inside surface to allow for the expansion. The first piece is always on the bottom. The J-channel will extend past the width of the frame and match up with the face of the vertical J-channel along the side. Measure the face of the J-channel and double it. Then, measure the width of the frame. Add these measurements together to determine the length of J-channel to cut.
2. Measure the desired length of J-channel on a piece of siding and use a square to mark the cut around the three surfaces of the channel.
3. Use the aviation snips to cut the piece of siding. Cut along the line on the face into the corner. Next, cut along the line on the nailing flange of the channel to the corner. Then bend the piece back over itself and slide the snips into the opening and cut the back to complete the cut.
4. The siding will need to be cut close to the frame of the opening when it is installed. For this to happen, cut a notch in the back of the J-channel to allow the siding to fit into place. Cut the notch from the end to match the width of the face of the channel, **Figure 3**.

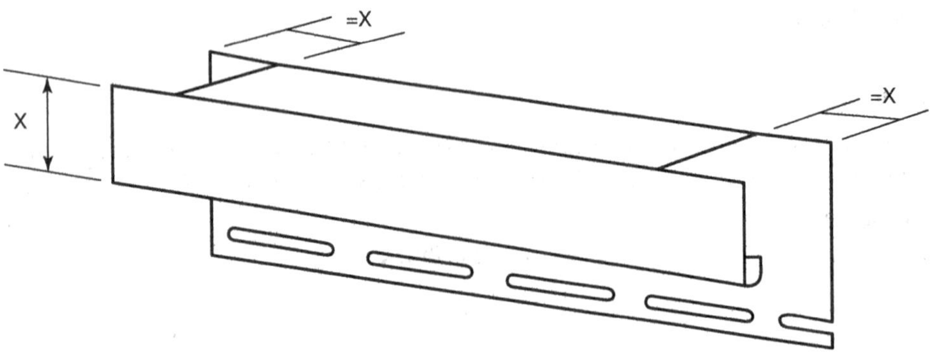

Goodheart-Willcox Publisher

Figure 3

Name ____________________

5. While holding the J-channel tight against the frame bottom, nail it in place. Be sure to stop the nail at the proper depth. **Figure 4** shows what this first step should look like.

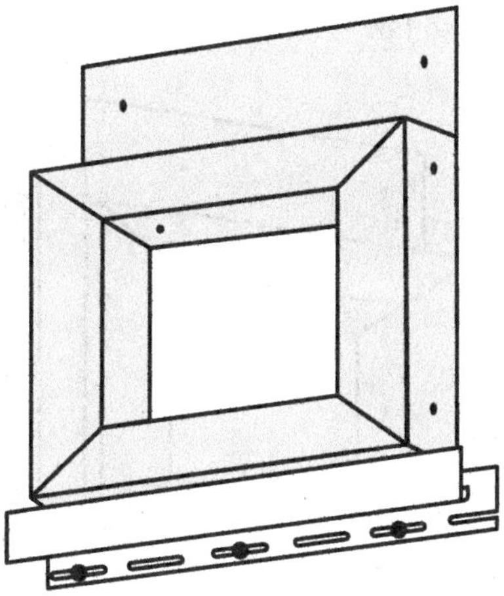

Goodheart-Willcox Publisher

Figure 4

6. Install the sides of the frame next. Measure the length of the frame side, and then add twice the width of the face of the J-channel to calculate the desired length. Lightly draw square lines on all three faces of the J-channel and cut using the same procedure as in Steps 3–4. Cut one for each side.
7. Prepare the right-side piece first. Hold the J-channel up in place to visualize where these cuts must be made. The top will have the notch cut out of the back just like Step 4 was done for the bottom piece. The notch should end up even with the top of the frame.
8. The bottom end of the J-channel will be cut at a 45° angle across the face to look like a mitered corner when complete. Use the square to lay out the 45° cut from the bottom outside corner. With the aviation snips, cut along that mark until it comes to the corner. Then, cut up along the corner to intersect the first cut and remove a small triangle of the J-channel face.
9. Again, the corner must be prepared for the siding to fit inside the J-channel. This time do not remove the notched piece of the back of the J-channel. Cut along the back corner equal to the width of the face and then bend the tab created back against itself to make a 90° flange. This tab will cover the inside corner so the same color will be visible inside the J-channel. **Figure 5** shows what the piece should look like.

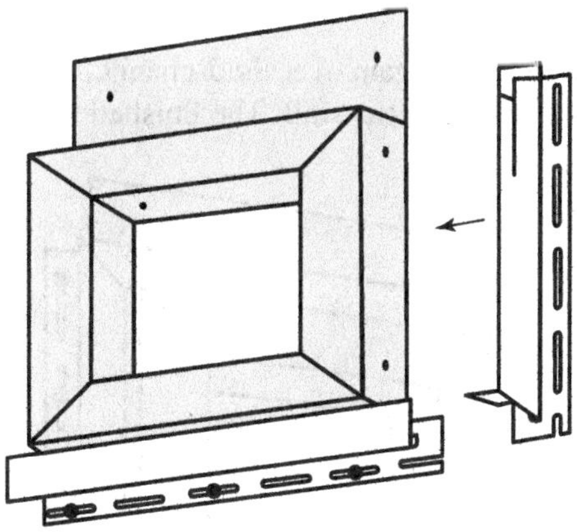

Goodheart-Willcox Publisher

Figure 5

10. Repeat Steps 7–9 for the left-side piece.

11. When fastening the piece to the wall, slide the angled end of the J-channel over and in front of the extension on the bottom piece. Hold it against the frame while nailing it to the wall. When nailing a vertical piece, always place a nail at the very top of the top nail slot so the piece hangs from that nail and does not slide down the wall. **Figure 6** shows what it should look like so far.

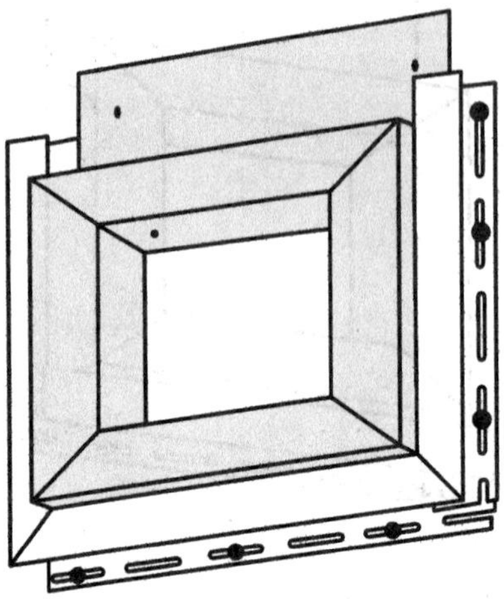

Goodheart-Willcox Publisher

Figure 6

12. Cut the top piece of J-channel to prepare it for the frame. This can now be measured from the outside edges of the faces of the J-channel on each side. Prepare this piece with a 45° cut on the face of both ends to complete the look of mitered corners. Cut both ends to form a tab to bend down over the ends of the frame to act like flashing protecting the frame from water, **Figure 7**.

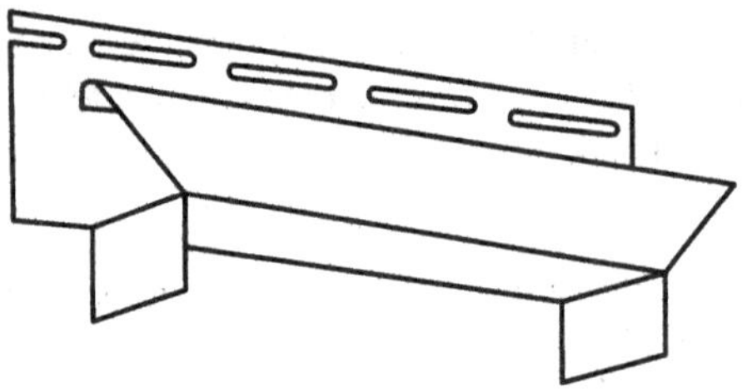

Goodheart-Willcox Publisher

Figure 7

13. Install this piece by sliding the mitered ends again over the J-channel side faces and in front of them. Hold the piece against the top of the frame and nail it to the wall. The finished frame should look like **Figure 8**.

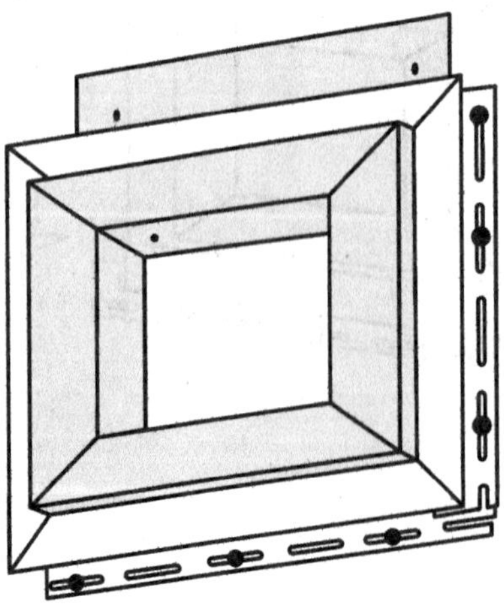

Goodheart-Willcox Publisher

Figure 8

Name ____

14. Before installing siding, fasten other trim pieces. At the bottom of a wall when a whole piece of siding will be used to start covering the wall, a starter strip is fastened at the desired height. Fasten a 36″ piece of starter strip on the wall 7″ below the frame. Use the 2′ level to make sure the strip is level when nailing it.

15. To the left of the frame and starting in line with the bottom of the starter strip, fasten a 31″ piece of J-channel vertically on the wall. The nailing flange of the J-channel should slide under the starter strip before nailing. Remember the top nail should be set at the very top of the highest nailing slot so the J-channel is hanging on it and holding it in position but can still expand and contract. Using the 2′ level, make sure it is plumb as it is nailed.

16. A 36″ piece of utility trim will be fastened 12″ above the frame to hold and secure the top of the final panel of siding. The left end of the trim should go into the J-channel when starting. Make sure the trim is level as it is nailed. The wall should look like **Figure 9**.

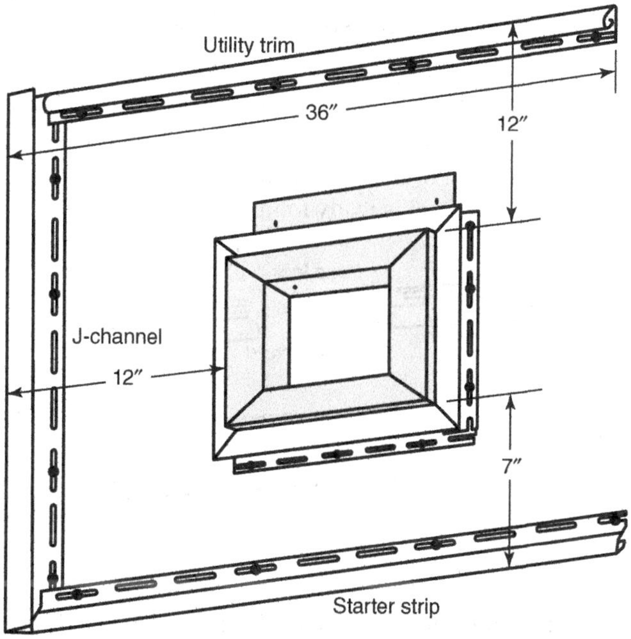

Goodheart-Willcox Publisher

Figure 9

17. You can now begin the installation of the siding. Measure a 36″ piece of siding from a larger panel. Draw a square line across the panel and cut it to length with the tin snips. When cutting across the center overlap and the top nailing flange, push the tin snips onto the piece as far as possible for the best leverage to make the cut.

18. Under a window or the eaves of a roof, lock the siding panel into a utility trim to hold it in place and give a finished appearance. Cut a piece of utility trim to the length of the bottom of the frame, slide it into the J-channel and tight to the bottom of the frame, and nail it in place on top of the J-channel nailing flange, **Figure 10**.

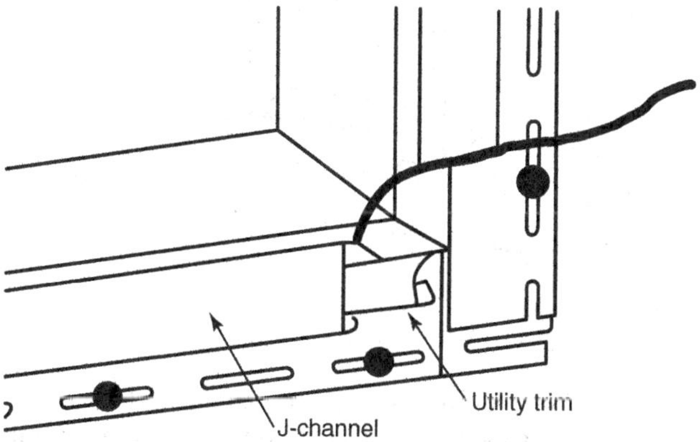

Goodheart-Willcox Publisher

Figure 10

19. Anytime the siding panel goes into a J-channel or corner post at the end, always keep it 1/4″ short to allow for the expansion and contraction of the panel. Hold the piece of siding on the wall with the bottom against the starter strip and 1/4″ from the inside of the J-channel. With the nailing strip on the frame, mark the siding 1/4″ wider on each side of the frame. With the square, draw a line a short distance down the siding from the top at each mark.
20. Measure from the bottom of the hook on the starter strip up into the J-channel to the top of the utility trim. Subtract 1/4″ from that measurement. From the bottom of the siding panel, measure up this distance and place a mark intersecting the lines drawn from the width of the frame. Use a straightedge to draw a line connecting these two points. This rectangle will be cut out to allow the siding to fit around the frame.
21. Using the tin snips, cut in from the top of the siding panel along the two lines at the sides of the frame until the cut reaches the horizontal line.
22. Place a straightedge on the horizontal line as a guide and cut a score line along it with the utility knife. The cut should not go through the panel. Then pick up the piece of siding and bend the rectangle at the cuts back against itself. Break the vinyl along the cut and remove it. You may need to bend it back and forth a couple of times get a complete break.
23. To hold this in place when installed, use the snap lock tool to create a small hook on the siding to hold it in the utility trim. About 2″ from the end of the cutout, slide the snap lock tool over the siding until it hits the built-in stop in the tool. Squeeze the handles together until the punch breaks through the siding. The hook created should be coming out the front of the siding panel. If not, turn the tool over and try again. Make a hook every 6″ along the cut edge. It should look like **Figure 11** and is ready to install.

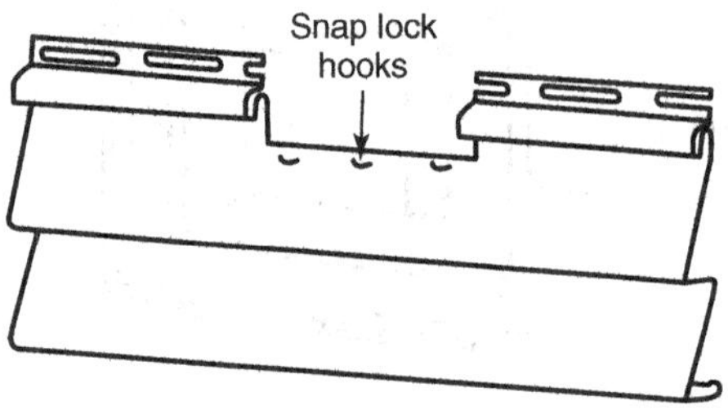

Goodheart-Willcox Publisher

Figure 11

24. Hold the siding panel just below the starter strip and line up the cutout with the frame. There should be a 1/4″ gap at the end inside the J-channel and a 1/4″ gap on each side of the frame. Slide the panel up onto the starter strip and guide the cut edge below the frame into the utility trim. The snap lock hooks should catch on the lip of the utility trim and hold the piece in place. Nail the piece in place without pulling up on the siding panel, **Figure 12**.

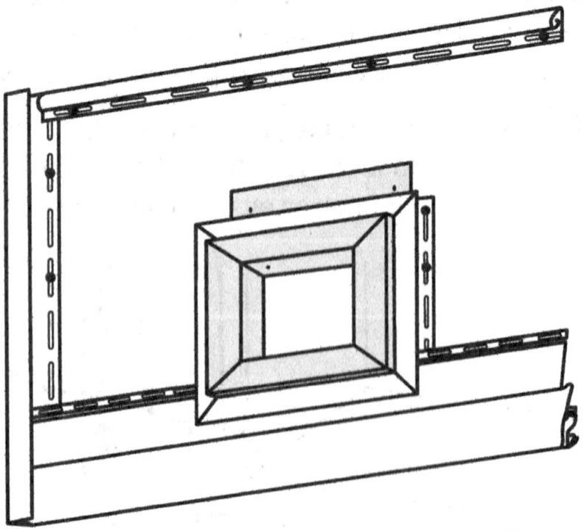

Goodheart-Willcox Publisher

Figure 12

25. Measure from inside the J-channel on the left to inside the J-channel at the frame. Subtract for a 1/4″ gap on each end for a total of 1/2″. Measure and cut a piece to fit. On the right, measure from inside the J-channel to the end of the piece below and subtract 1/4″ for expansion. Measure and cut a piece to fit.

Name ______________________________

26. On the left side of the frame, above the frame, insert the left end of the siding piece into the J-channel. Slide the piece down inside the J-channel on both ends until it is below the lock strip on the siding piece below. The J-channel may have to be bent out a little bit to get the siding to pass over the nailing hem on the piece below. Push the siding tight against the wall and slide it up so it locks into the lower piece. Check to see it is centered side to side in the space and then nail.

27. On the right side of the frame, hold the siding piece below the lock and slide it into the J-channel leaving the 1/4″ gap and push it up to lock into the piece below. Nail it in place. Refer to **Figure 13**.

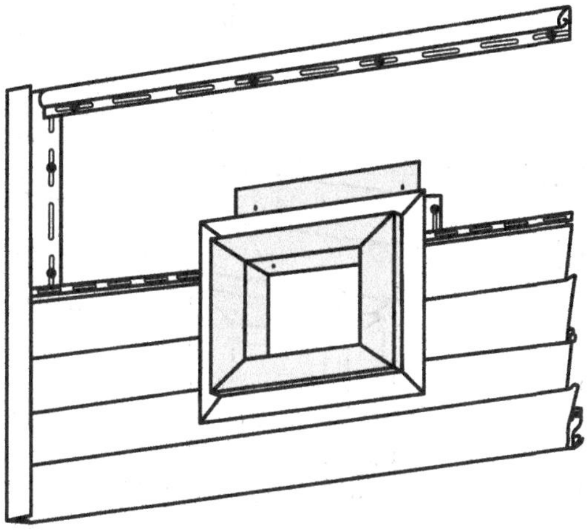

Goodheart-Willcox Publisher

Figure 13

28. Cut another 36″ piece of siding. Hold it above the frame and slide it into the J-channel to the left leaving the 1/4″ for expansion. Place a mark 1/4″ away from either side of the edges of the frame like it was done on the piece below the frame.

29. Next, measure for the height of the cutout. Hook the tape measure on the lock of the panel below and measure up to the top of the frame. Add the 1/4″ to allow for expansion. Lay out this measurement from the bottom of the panel at the marks for the width of the frame. Lay out, cut, and break out this rectangle for the frame the same way it was done in Steps 21–22 for panel below the frame.

30. To install this panel, place it above the frame and slide it down into the J-channel on each side. When it is all the way down, slide the siding removal tool under the bottom edge of the panel near one side of the frame and hook the locking lip. Pull the lip down past the lower lock and twist the tool up and out from under the lip while putting some pressure on the bottom edge of the panel. The locking hook should engage and connect the two panels together. Do the same for the other side of the frame to lock that side. Nail the panel in place. Compare the work to **Figure 14**.

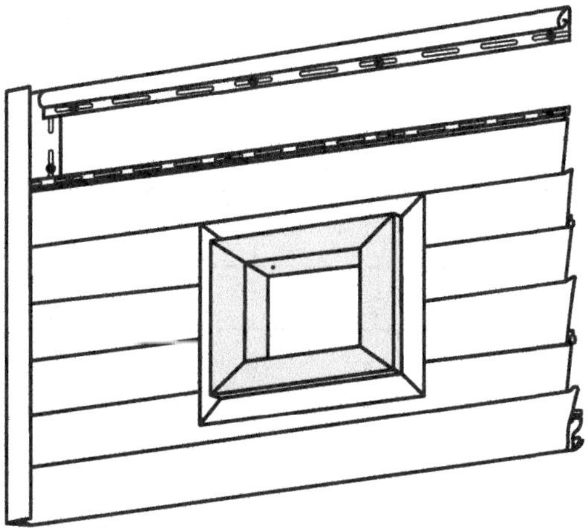

Goodheart-Willcox Publisher

Figure 14

31. Secure the final piece in place by hooking it into the utility trim just like below the frame. Measure from the panel lock on the last piece up to the top of the utility trim and subtract 1/4″. Check the measurement at each end of the panel and cut the piece to fit if needed. On another piece of siding 36″ long, measure up from the bottom of the panel on each end and mark. Using a straightedge and a utility knife, score and break the top nailing flange off the panel at the measurement marked.

32. Use the snap lock punch tool to make the lock punch every 6″ along the top of the panel. Be sure the punch is going out the front of the panel. Hold the siding panel in place below the lock and slide it up into place, making sure the top edge slides up into the utility trim and locks in place. Check the placement of the panel so there is a 1/4″ space inside the J-channel at the end. The finished product should look like **Figure 15**.

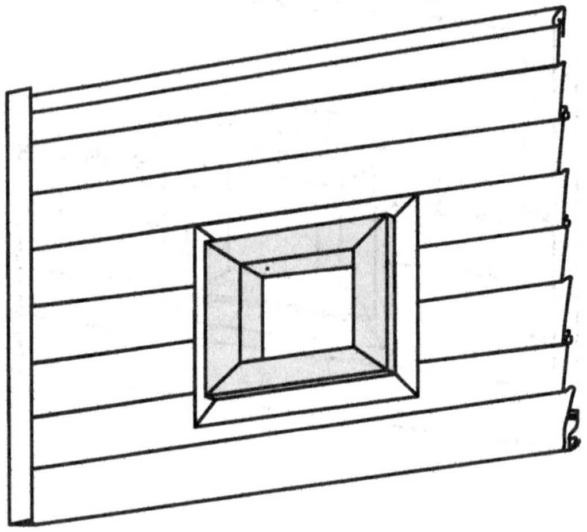

Goodheart-Willcox Publisher

Figure 15

Review and Analysis

1. When cutting the J-channel for the top of the frame, why is the tab created on the ends and bent down over the edges of the frame?

____

____

____

2. Why is the 1/4″ gap at the end of the panel when the siding goes into the J-channel so important?

____

____

____

3. What will happen to the siding panel if it is nailed too tightly to the wall?

____

____

____

Name ______________________ Date ____________ Class ____________

CHAPTER 6 MEP Systems Rough-In

Activity 6-1: Chapter 6 Review

Answer the following questions after studying the chapter.

Lesson 6.1

1. _____ Each year, the National Association of Home Builders sponsors the _____, the largest event of its kind, offering educational, networking, and product showcase opportunities for all segments of the residential construction industry. (6.1-1)
   A. Equip Exposition
   B. International Builders' Show
   C. Greenbuild International Conference and Expo
   D. International Surface Event

2. _____ *True or False?* The National Association of Home Builders is closely affiliated with the Republican Party. (6.1-1)

3. _____ The Home Builders Institute is dedicated to _____. (6.1-1)
   A. creating vocational training initiatives
   B. developing industry safety standards
   C. promoting green construction
   D. updating building codes

4. _____ *True or False?* Trade unions are more significantly represented in residential construction than in commercial construction. (6.1-1)

Lesson 6.2

1. _____ *True or False?* Credentials can include both professional and academic qualifications. (6.2-1)

2. _____ Trade certifications are typically issued by _____. (6.2-1)
   A. community colleges
   B. professional organizations
   C. your construction industry peers
   D. an owner of a construction company

3. _____ *True or False?* Some certifications require a minimum level of education, such as online training. (6.2-1)

4. _____ HVAC technicians that want to work with equipment that could release refrigerants into the atmosphere must be certified by which government agency? (6.2-1)
   A. EPA
   B. CDC
   C. OSHA
   D. FEMA

Lesson 6.3

1. _____ A loop or continuous path from the source of electric power, through various loads, and then back to the source is called a _____. (6.3-1)
   A. conductor
   B. junction
   C. circuit
   D. receptacle

2. _____ *True or False?* To install rough plumbing, plumbers connect pipes to the face of drywall. (6.3-1)

3. _____ *True or False?* When installing plumbing runs below the first floor of a residence, each run must be installed through the joists. (6.3-1)

4. _____ Today, most supply-side plumbing is installed using _____. (6.3-1)
   A. copper tubing
   B. PVC
   C. PEX
   D. CPVC

5. _____ Heat moving through a metal spoon placed in a bowl of hot soup is an example of _____. (6.3-2)
   A. radiation
   B. reflection
   C. convection
   D. conduction

6. _____ Which type of heating system uses the resistance of conductors to produce heat? (6.3-2)
   A. Boiler
   B. Traditional stove
   C. Heat pump
   D. Electric furnace

7. _____ Which type of heating system might be installed under subflooring with insulating beneath the heating tubes? (6.3-2)
   A. Hydronic radiant heating system
   B. Direct heating system
   C. Hydronic perimeter heating system
   D. Forced-air perimeter system

8. _____ What is the main disadvantage of electric resistance baseboard heating systems? (6.3-2)
   A. They are difficult to install.
   B. They are expensive to operate.
   C. They require a fuel storage container.
   D. They are expensive to purchase.

9. _____ *True or False?* Residential air-cooling systems often use the same air distribution system as the heating system. (6.3-2)

Lesson 6.4

1. _____ Electricians typically install _____ during electrical rough-in. (6.4-1)
   A. light switches
   B. appliances
   C. receptacles
   D. conductors

2. _____ A device that protects conductors from overload by shutting off power to the circuit if an excessive current draw occurs is a _____. (6.4-1)
   A. conduit
   B. connector
   C. circuit breaker
   D. service panel

3. _____ *True or False?* In electrical systems that use fuses, if too much current flows through the circuit, the fuse closes the circuit. (6.4-1)

4. _____ Which weights of copper tubing are typically used for water supply lines? (6.4-1)
   A. A, B, and C
   B. F, G, and H
   C. K, L, and M
   D. N, P, and R

5. _____ *True or False?* PEX tubing is only used for cold water supply lines because PEX cannot stand up to the heat of most residential hot water systems. (6.4-1)

6. _____ Which statement about tankless water heaters is true? (6.4-1)
   A. The water that enters the unit is preheated by a boiler.
   B. They are much larger than tank-style water heaters.
   C. All types require a propane connection for installation.
   D. They contain a heat exchanger that rapidly heats the water.

7. _____ *True or False?* Before a masonry anchor for HVAC duct work can be installed, a pilot hole must first be drilled. (6.4-2)

8. _____ *True or False?* When using pipe dope, the tape is wrapped counterclockwise around the male threads before the connection is made. (6.4-2)

Name ____________________

Lesson 6.5

1. _____ *True or False?* This tool is used to bend conduit. (6.5-1)

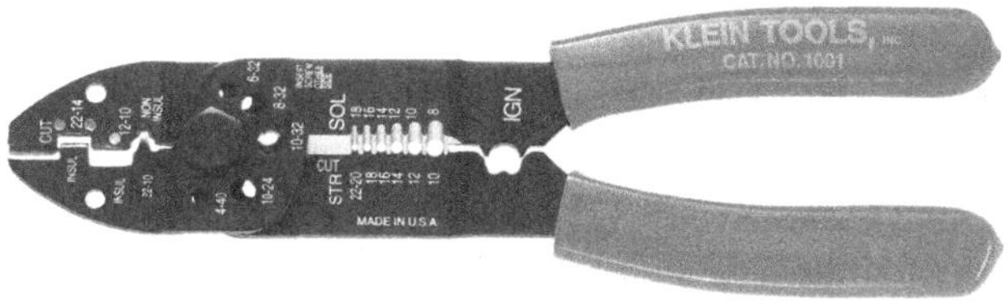

Klein Tools

2. _____ *True or False?* Pliers are good for shaping wire. (6.5-1)

3. _____ Which tool is used to pull wire through conduit? (6.5-1)
   A. Fish tape
   B. Multipurpose tool
   C. Slip-joint pliers
   D. Snips

4. _____ Boring tools are used for _____. (6.5-1)
   A. cutting threads on steel pipe
   B. drilling through masonry materials
   C. soldering copper pipes and fittings
   D. making large holes in wood framing members

5. _____ Which tool is used to remove the burrs left inside of a pipe that has been cut? (6.5-1)
   A. Reamer
   B. Die
   C. Chain wrench
   D. Torch

Lesson 6.6

1. _____ *True or False?* Using an undersized extension cord with a power tool or appliance can cause an electrical fire. (6.6-1)

2. _____ *True or False?* An arc flash is quick discharge of electrical energy through the air caused by a high-voltage gap and a breakdown between conductors. (6.6-1)

3. _____ What is a common cause of electrical incidents? (6.6-1)
   A. Excessive ground-fault protection
   B. Using battery-operated tools
   C. Contact with power lines
   D. Working with a stable ground connection

4. _____ *True or False?* The ground connection provides a path of greatest resistance that forces the current to return to the source and complete the circuit. (6.6-1)

5. _____ *True or False?* Extension cords used in construction should have three wires. (6.6-1)

6. _____ *True or False?* Lockout is the practice of placing a tag on a mechanism or electrical switch to inform others that service work is in process. (6.6-1)

Lesson 6.7

1. _____ What is the area of a rectangle that is 20″ long and 15″ wide? (6.7-1)
   A. 62.832 in^2
   B. 150 in^2
   C. 300 in^2
   D. 471.24 in^2

2. _____ What is the area of a triangle that has a base of 16′ and a height of 7.5′? (6.7-1)
   A. 60 ft^2
   B. 94.248 ft^2
   C. 120 ft^2
   D. 188.496 ft^2

3. _____ *True or False?* The total distance across a circle through its center point is its radius. (6.7-1)

4. _____ A circle with a diameter of 9′ has a circumference of _____. (6.7-1)
A. 14.1372 ft
B. 28.2744 ft
C. 63.6174 ft
D. 254.4696 ft

5. _____ A circle with a diameter of 10″ has an area of _____. (6.7-1)
A. 31.416 in^2
B. 78.54 in^2
C. 157.08 in^2
D. 314.16 in^2

6. _____ What is the area of a half-circle with a radius of 22″? (6.7-1)
A. 69.1152 in^2
B. 380.1336 in^2
C. 760.2672 in^2
D. 1520.5344 in^2

Lesson 6.8

1. _____ *True or False?* All electrical print symbols are standardized to clearly communicate the house's wiring and fixtures. (6.8-1)

2. _____ Information associated with the distribution panel, including the number and sizes of circuit breakers and the devices protected by each one, is provided in a(n) _____ schedule. (6.8-1)
A. distribution
B. equipment
C. lighting
D. panel

3. _____ On a plumbing plan, what does this symbol represent? (6.8-1)

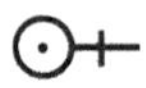

A. Elbow-turned up
B. Elbow-turned down
C. Reducer-concentric
D. Reducer-offset

Lesson 6.9

1. _____ Traditional furnaces and boilers _____. (6.9-1)
A. operate at full capacity at all times
B. operate at full capacity until the heating demand stops
C. adjust their heat output based on the desired indoor temperature
D. adjust their heat output based on the actual outdoor temperature

2. _____ *True or False?* Condensing boilers compare outdoor air temperature with indoor air temperature and adjust their output using modulating control systems. (6.9-1)

3. _____ Using refrigeration principles, a(n) _____ transfers heat from one place to another. (6.9-1)
A. electric stove
B. heat pump
C. direct heating system
D. forced-air perimeter system

4. _____ *True or False?* The amount of electrical energy consumed by a heat pump is greater than the amount of heat the pump generates. (6.9-1)

5. _____ Phantom power consumption commonly occurs when a _____. (6.9-1)
A. television is in standby mode
B. lamp is unplugged
C. rechargeable device is charging
D. window-mount air conditioner is running

Name ______________________ Date ____________ Class ____________

Activity 6-2: Project—Roughing-In Electrical Wiring

Learning Outcomes

After completing this lab activity, you will be able to:

- Demonstrate how to install wires between device boxes in the rough-in stage.
- Demonstrate how to prepare wires before connecting to a device.
- Identify the hot, neutral, and ground wires in the system.

Introduction

In this exercise you will rough wire in the electrical needs for the installation of a duplex outlet, light switch, and a light. Another important skill to have is the ability to fix an extension cord or cord to an electrical appliance. An electrical cord will need to be wired to provide power to the model. The *National Electric Code (NEC)* contains the rules that must be followed when wiring a home just like the IRC is followed to build the home.

Text Reference

Chapter 6, *MEP Systems Rough-In*

Equipment

The following tools and equipment are needed to perform these activities:

- #2 Phillips screwdriver
- 14′ of 12/2 with ground Nonmetallic (NM) cable, commonly known as Romex wire
- Cable staples, 9/16″
- Carpenter's pencil
- Chalk line
- Cordless drill with #2 Phillips bit with magnetic extension
- Extension cord, 6′
- Heavy duty 1/2″ drill or right-angle drill
- Lineman pliers
- Measuring tape
- NM cable ripper
- Round PVC light box
- Safety glasses
- Spade bit, 5/8″
- Utility knife
- Wire strippers

Safety Review

Follow all safety procedures specified by your instructor and specified in any appropriate manufacturer's information. The following are some additional safety-related items to keep in mind:

- Wear safety glasses at all times.
- Consider all wires as "HOT" until circuits have been turned off or proper testing has been done to verify.
- All metal can have sharp edges or corners; be careful not to cut yourself.
- Ensure extension cords are properly grounded.
- Hold the drill securely while drilling holes for the wires.

Procedure

Power will be provided to this electrical model by wiring an electrical plug that can be plugged in to check that the duplex outlet and light work and that the wiring has been performed correctly. Locate the electrical boxes first and then drill holes for the Romex. The White wire is neutral, the Green or bare copper wire is for ground and the Black wire is the Hot wire or the feed to the circuit. Each fixture is usually connected to the ground wire for safety in the system to protect users from getting a shock.

Review Floor Plan for Placement of Electrical Fixtures

1. On a stud on the left side of the construction model, attach one single gang box 10″ from the bottom plate using the nails provided with the box. Place the box so the little bumps on the side are against the edge of the 2×4 stud. This should set it so 1/2″ of the box is out past the edge of the stud and will be even with the drywall when completed.
2. Attach the other single gang box 24″ up from the plate on the stud in the right corner of the construction model. It should also extend 1/2″ past the edge of the stud to allow for the installation of drywall.
3. Attach the round box for the light near the end of the short wall on the inside of the top plates. See **Figure 1** for completed box placement.

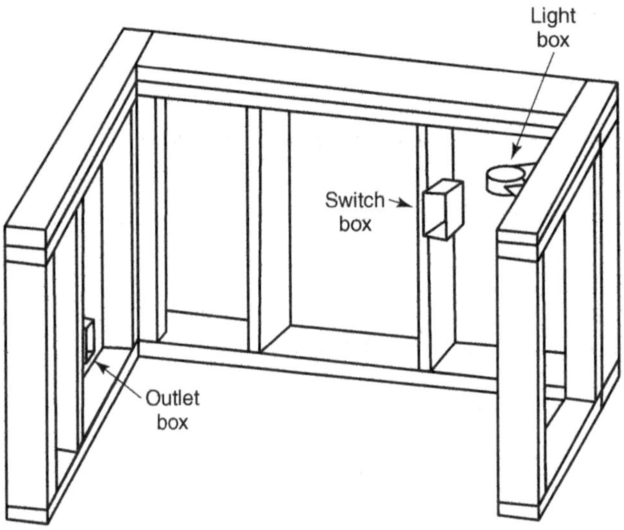

Goodheart-Willcox Publisher

Figure 1

4. Mark studs 16″ up from the bottom plate to locate the holes for the NM wire. Snapping a chalk line is the fastest way to mark where 16″ is along the length of a wall.
5. Tighten the 5/8″ spade bit into the chuck of the drill. Drill holes for the NM through each 2×4 and 2×6 stud between the single gang boxes. The hole should be in the center of the width of the stud to ensure a drywall screw will not penetrate the wire when hanging the drywall. At the left corner of the model, drill at a 45° angle through the center of the studs to allow the wire to easily make it around the corner.
6. Drill another hole straight up from the light switch box through the double top plates.
7. From the left side of the model, starting at the lower outlet box, feed the wire through the holes in the studs until it reaches the other side and at the switch box.
8. Remove about 9″ of the plastic sheathing on the NM wire to expose the conductors inside. Use the cable ripper for this by sliding the wire between the wings of the ripper and through the hole in the wide end. Squeeze the point inside the ripper through the cable covering and slide it to the end of the wire, slicing the plastic covering. Peel the plastic to one side, separating it from the wires, and cut it off with a utility knife close to where the slit starts.

Name ______________________________

9. Use a screwdriver or pliers to break out one of the plastic tabs in the lower back of the switch box to feed the wire into the box. Thread the wire into the box from the back side until the plastic sheathing is just inside the box. There should be at least 6″ of wire coming out past the front edge of the box during rough-in steps.
10. The *NEC* states that the NM cable must be secured to the framing within 8″ of the box in a case like this. Neatly bend the wire below the box to the stud and secure it with a cable staple about 4″ below the box. See **Figure 2**.

Goodheart-Willcox Publisher

Figure 2

11. Back at the first box, hold the wire next to the box and determine where to cut it. Again, there should be about 6″ of wire coming out past the front of the box for this stage of rough-in. Cut the wire with the lineman pliers. Remove the sheathing again as before and thread the wire into the box from the top. Secure the cable to the stud with a staple about 4″ from the box again.
12. Prepare one end of the remaining cable to go into the top of the switch box. Thread it into the box from the top and secure it to the stud as the others. Run it to the round light box by feeding the cable up through the hole in top plates and loop it over to the round light box.
13. Cut the cable to go into the box with the 6″ of wire past the front of the box. Strip the sheathing off, insert it into the box, and secure it on the top of the plate with a staple.
14. The remaining wire will be used to provide the connection to a power source used to energize and check the completed wiring job. Prepare one end of the wire, insert it into the bottom of the outlet box on the left side of the unit, and secure it below the box to the stud. See **Figure 3** showing cable runs.

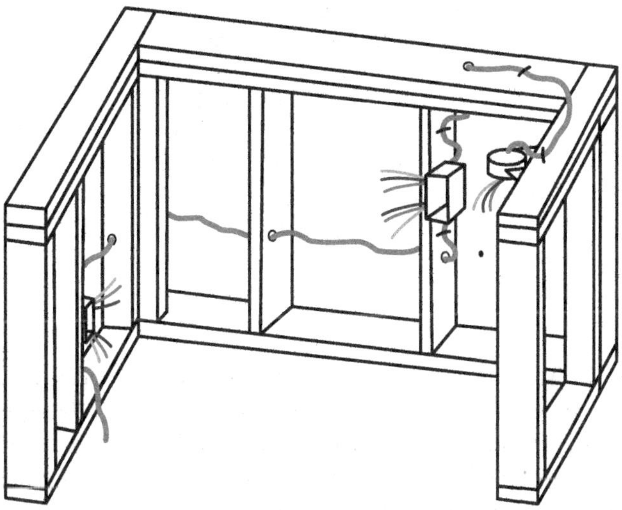

Goodheart-Willcox Publisher

Figure 3

15. Using the wire strippers, remove about 3/4″ of insulation off the ends of all wires coming out of boxes. There are a series of holes at the end of the wire stripper tool based on the size of the wire being worked on. The conductors are size 12 in the wire being used for the unit. Using that hole, put the wire in the strippers at the hole labeled 12 and clamp it tight. Loosen the grip on the wire just a little and with a little twist, push the insulation off the wire to expose the conductor. See **Figure 4**.

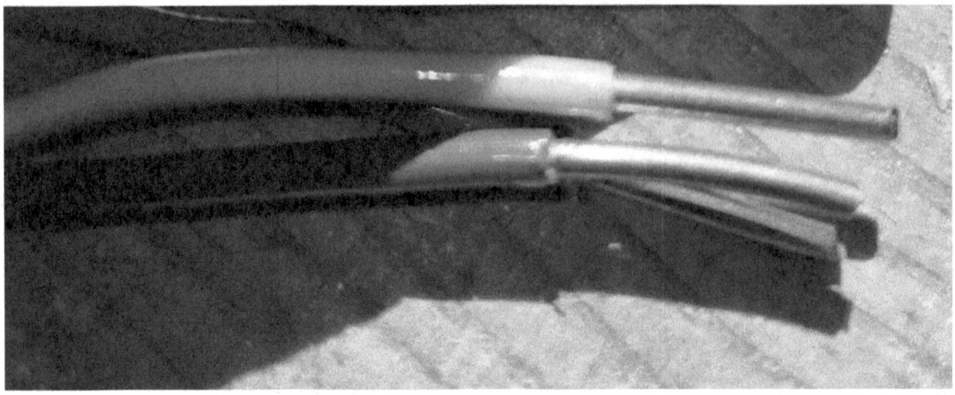

Goodheart-Willcox Publisher

Figure 4

16. Roll all the exposed wires into a circle or partial circle and push them back into the electrical boxes so they are protected while the wall finishes are being installed. During the trim out or finish stages of the construction process, the wires will be pulled back out and connected to the devices to operate the electricity in the home. See **Figure 5**.

Goodheart-Willcox Publisher

Figure 5

17. Cut the female end off the extension cord and remove a few inches of the plastic covering to expose the conductors. Do the same on the end of the short NM cable coming from the bottom of the outlet box. Strip some of the insulation off each conductor and twist the wires together matching the colors in pairs. Secure a wire nut on each pair of wires and wrap each with some electrical tape. When the wiring in the unit is complete, this cord will be plugged into an outlet to energize the wiring to check the work.

Name ____

Review and Analysis

1. What other operations can be accomplished with the wire stripper during the wiring process?

____

____

____

2. What are the basic colors of electrical wires and what do they represent?

____

____

____

3. Why do we have a ground wire when we wire electrical fixtures or appliances?

____

____

____

Notes

Name ______________________ Date ______________ Class ____________

Activity 6-3: Project—Roughing-In Plumbing

Learning Outcomes

After completing this lab activity, you will be able to:

- Demonstrate how to cut copper pipe.
- Demonstrate how to cut PVC pipe.
- Identify the pipe and fittings needed to drain and vent a plumbing fixture.
- Explain how to solder a copper pipe joint.
- Demonstrate how to fasten pipe to blocking.

Introduction

The plumbing rough-in is completed while the walls of the structure are still unfinished. This allows builders to run the pipes where needed without any wall finishes in the way. Most plumbing fixtures need both water supply lines and a drain. The drain needs to be connected to the main drain that will lead all wastewater away from the house. This drain also needs to be vented to allow water to flow smoothly. The materials used for roughing in will be PVC pipe for the DWV (drain, waste, and vent) system and copper pipe for hot and cold water supply.

Text Reference

Chapter 6, *MEP Systems Rough-In*

Equipment

The following tools and equipment are needed to perform these activities:

- 100 grit wet and dry sandpaper or strip of emery paper
- 3d box nails
- 16d common nails or green vinyl sinkers
- 48″ of 2×4 lumber
- Carpenter's pencil
- Circular saw or power miter saw
- Copper 90° elbows (4), 1/2″
- Copper end caps, 1/2″
- Copper pipe, 10′ of 1/2″
- Copper pipe straps (2), 1/2″—two hole or equivalent
- Copper stubout brackets (8), 20″ of 1/2″
- Deburring tool
- Fitting brush for copper pipe and fittings
- Flat head or pan head screws, 1 1/4″ long
- Flux brush
- Flux, non-corrosive
- Hacksaw
- Hammer
- Hanger strap, 20 inches of 3/4″
- Heat absorbent pad or fire blocking shield
- Heavy-duty drill or right-angle drill, 1/2″
- Measuring tape
- Multispur bit, 2″
- Pipe cutter
- Pipe isolator inserts (4), 1/2″
- Plumbers drill bit, multispur bit, auger bit, or hole saw, 1 3/8″
- Propane
- Propane torch
- PVC 90° elbows (2), 1 1/2″
- PVC cement
- PVC pipe, 6′ of 1 1/2″
- PVC primer
- PVC sanitary vent tee, 1 1/2″ × 1 1/2″ × 1 1/5″
- Reamer, to remove burrs
- Safety glasses
- Single cut file with a handle
- Solder, lead-free 1/8″ solid core
- Spark lighter or striker

Safety Review

Follow all safety procedures specified by your instructor and specified in any appropriate manufacturer's information. The following are some additional safety-related items to keep in mind:

- Wear safety glasses at all times.
- When cutting materials, always cut away from your body.
- Wear gloves when working with hot items.
- Be aware of flammables when soldering.

Procedure

When installing plumbing fixtures, there will be a drainpipe coming up from the foundation or basement attached to the main sewer line. From this drainpipe, a horizontal drain line attaches to each of the fixtures throughout the house. The sink for this activity will be located in the center of the wall. There will be two water lines: one for cold water and one for hot water, as well as a drain line. Use **Figure 1** for reference as the plumbing is installed in this project. The PVC will be fastened together using PVC primer and cement, and the copper pipe will be soldered.

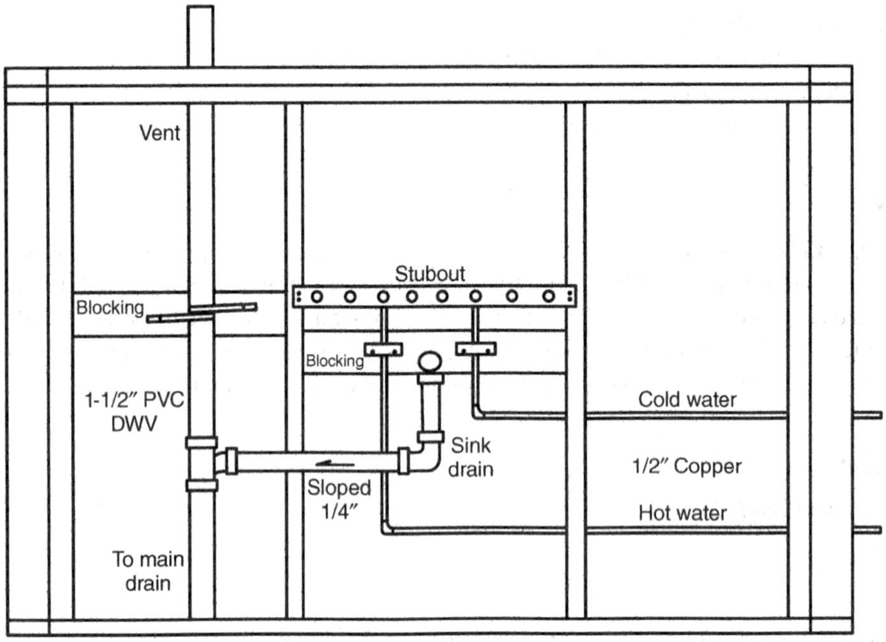

Goodheart-Willcox Publisher

Figure 1

Drain Cutting and Fitting

1. Facing the construction model, on the left side, measure and mark 7″ from the first stud back toward the corner. Mark the center of the plate at the 7″ mark to locate the center of the drain hole. Drill a 2″ hole to a depth of 1 1/4″. This is where the drain line will come up into the house. See **Figure 2**.

Goodheart-Willcox Publisher

Figure 2

2. Repeat the same measurements on the bottom of the top plate to locate the hole for the continuous vent going up into the house.

3. Drill upward through top plate with the 2″ drill bit. This will be for the vent pipe.

> **NOTE**
> The vent pipe is located to the side of the sink because the electrician will need room to install a light fixture above the sink.

4. Take the 1 1/2″ PVC pipe and cut a piece 8″ long. Use a hacksaw or 1 1/2″ PVC cutter.

5. Use sandpaper, file, or deburring tool to remove any burrs on the edges of the cut end of the pipe to make the cut smooth. By smoothing the cut, the wastewater will flow freely through the pipes and fittings.

> **NOTE**
> This will be the standard procedure each time a piece of PVC pipe is cut.

6. Insert the 1 1/2″ × 1 1/2″ × 1 1/2″ sanitary vent tee onto the 7″ pipe. Do not secure with pipe cement yet. Leave it loose.

7. Cut a 46″ long piece of 1 1/2″ PVC pipe.

8. Slide the cut PVC through the top plates down into the top of the sanitary tee vent. Do not secure with glue.

9. Measure the height from the bottom plate to the center of the side opening of the sanitary vent tee. Mark this measurement with an additional 1/8″ on the stud to the right of the tee. This will allow for a small slope to properly drain water from the sink. To help make this measurement, there are little PVC bumps or tick marks around the end of the fitting, called the hub. These tick marks help line up a fitting with one another while gluing. There will be a tick at the center of the end to measure from to accurately locate the hole in the stud.

10. At 2″ in from the face of the stud, drill a 2″ hole through the stud, aiming for the center of the sanitary vent tee.

11. Cut a piece of 1 1/2″ PVC pipe so it extends from the sanitary vent tee to the center of the wall where the sink would be located. There will be another 90° elbow making a turn up to the sink drain that must be accounted for when making this measurement. Measure from inside the hub of the tee to the center location of the sink on the wall. From this measurement, subtract the fitting allowance from the fitting as shown in **Figure 3**. This is the length to cut the pipe to center it where desired.

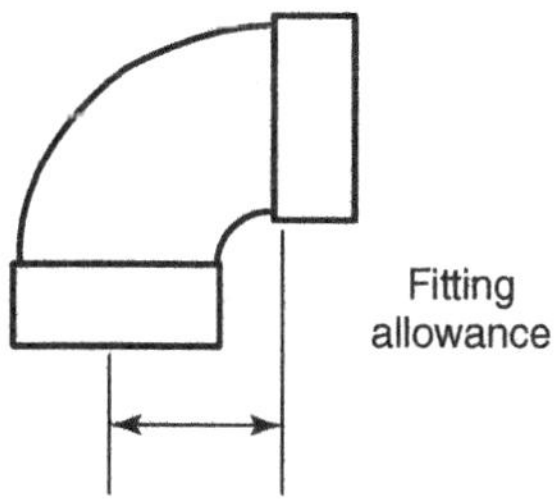

Goodheart-Willcox Publisher

Figure 3

12. Cut another piece of 1 1/2″ PVC pipe that will go up to the rough-in height of the sink, 16″ above the bottom plate in the wall. Just like in the previous step, allow for the height of the fitting to get the actual length of pipe to cut. Dry fit these together for now.

13. Add a piece of 2×4 blocking behind the vent line. Nail in with two 16d nails on each side. A strap to hold the pipe in place will be added later. See **Figure 4** showing the waste system fully prepared.

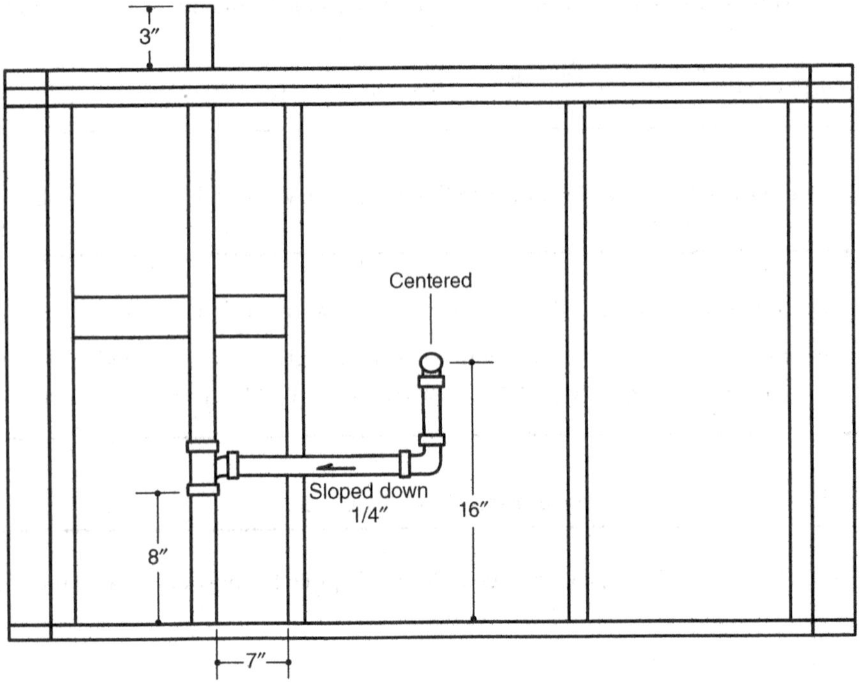

Goodheart-Willcox Publisher

Figure 4

Water Line Cutting and Fitting

14. On the right side of the model, measure 5 1/2″ and 9 1/2″ up from the bottom plate and mark the two studs with these measurements near the back side of the stud. This will be the location of the hot and cold water lines.
15. At the 5 1/2″ and 9 1/2″ marks, measure 2″ inward from the outside of the stud. These measurements are the center lines for the 1 3/8″ holes needed for the isolator inserts.
16. Drill 3/8″ holes at each of these locations.
17. Insert the 1/2″ pipe isolator fittings in these holes. These will keep the pipes from chattering and bouncing in the wall when water is running. Tap them gently in with a hammer. Do not nail them in yet. See **Figure 5**.

Goodheart-Willcox Publisher

Figure 5

Name ____

18. On the center two studs in the wall, measure up 21″ to locate and fasten the copper stubout bracket. Center the 1/2″ holes 4″ to each side of the drain and fasten the bracket to the studs. This will also be 5″ above the drainpipe opening, center to center.

19. Measure two pieces of copper pipe to run from the back corner of the unit to the stubout bracket behind the sink. Measure one to align with one 1/2″ opening on the stubout bracket, 4″ to the left of the center drain. Measure the second one to align with an opening in the stubout bracket, 4″ to the right of center. Add an extra 3″ and cut to length with the pipe cutter.

NOTE

Use the pipe cutter to cut the copper pipe. Unscrew the knob on the cutter until it fits over the pipe. Slide the pipe in and line up the measurement mark with the cutting disc in the center of the cutter. Tighten the knob until the cutting wheel contacts the pipe with just a slight amount of pressure. Spin the pipe cutter around the pipe a couple of times and then tighten the knob a little bit more. Repeat this spinning and tightening process until the cutter breaks through the copper.

20. Deburr the pipe pieces by cutting with the deburring blade on the pipe cutter or using a reaming tool. Removing the burr is a process that will be completed on all the copper pipe after cutting. The pipe cutter pushes some of the copper inside the pipe as it cuts, creating the burr that will impede the flow of water through the pipe.

NOTE

Check all fittings for burrs as well. If burrs are present and not removed, the pipe will not fit properly inside to make a good solder joint.

21. Fit the longer 1/2″ copper pipe in the bottom holes with the isolator inserts and push in to align with the 1/2″ hole, 4″ to the left side of the drain on the stubout bracket.

22. Repeat Step 21 for the shorter pipe in the upper holes with isolator inserts.

23. Align the end of pipe with the 1/2″ hole on the right side of the stubout bracket, 4″ from the drain.

24. Add 1/2″ 90° copper pipe elbows to the two pipes below the stubout bracket.

25. Cut two pieces of 1/2″ copper pipe to go from the elbows just installed running up to the stubout bracket. Allow for the space the 90° elbow uses just like was done for the PVC drainpipe.

NOTE

To aid with accuracy, place an elbow on a short piece of pipe and hold it in position with the pipe through the stubout bracket and the elbow above the other. Measure from one elbow to the other to obtain the length of pipe needed. Be sure to include the pipe length going into the fitting.

26. Insert each of these pipes into the elbow below up to the stubout height. Then add another elbow on the end of these pipes turning out to the face of the wall.

27. Cut two pieces of 1/2″ copper pipe to go through the stubout bracket to fit into the 90° elbow on the vertical pipe pieces. Measure this length with the vertical pieces held plumb in the wall and long enough so 4″ of pipe extends past the studs to allow for the finish plumbing to be completed in the future.

Drain Final Assembly

28. Disassemble the vent and drain lines making sure there are no burrs around the cut ends and the parts are clean. Mark how they fit together for reassembly with the PVC cement.

29. Apply PVC primer to the inside of the hub at the center opening of the sanitary vent tee and to the end of the piece of pipe leading to the sink. Be careful not to drip over other things while using the primer. Allow the primer to work for a couple of minutes but no more than five; then, apply the PVC cement to the hub and the end of the pipe in the same area as the primer. Push them together and twist them about a quarter of a turn. Place this pipe through its hole in the wall ready for the next step.

30. At the main drain, apply PVC primer and then PVC cement to the first 7″ pipe coming up from the wall plate and to the bottom hub of sanitary vent tee. Push them together with a twisting motion.

> **NOTE**
> Repeat the steps of preparing the joints with PVC primer and then applying the PVC cement at each step of joining fittings and pipe in a drain system. The PVC primer cleans and softens the PVC so the cement can adhere properly to the pipe and fittings. The PVC cement creates a chemical reaction that melts the PVC between the pipe and fitting to form the bond. This slight melting of the PVC is referred to as a cold weld. The twisting of the parts during assembly ensures the cement is spread around the joint evenly.

31. Assemble the remainder of the vent to the top opening of the sanitary vent tee.
32. Assemble the rest of fittings and the pipe that go up for the sink drain.
33. Cut an 18″ piece of hanger strap, wrap it around the vent pipe at the blocking, and attach it to the blocking with two short screws. See **Figure 6**.

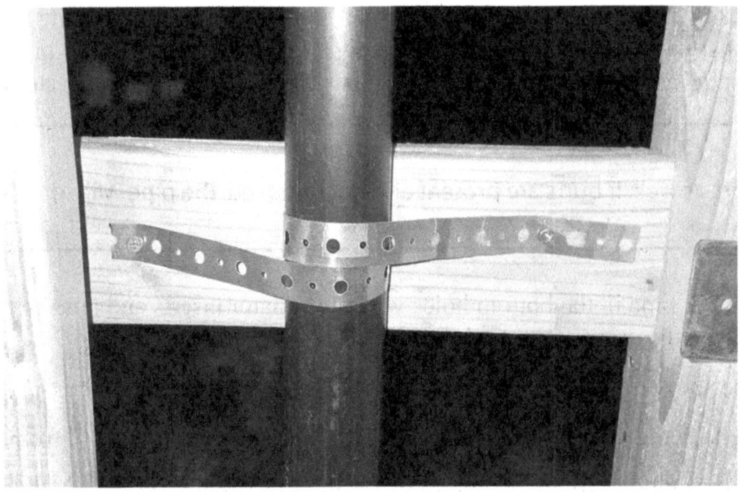

Goodheart-Willcox Publisher

Figure 6

Water Line Final Assembly

34. Disassemble the copper pipe and fittings one water line at a time.
35. Starting at the bottom, clean all pipe joints and fittings. Do not touch the cleaned areas with your hands as that will leave oils from your hands that will affect the bonding of the solder.

> **NOTE**
> One of the most important steps in achieving leak-free joints using copper piping is the cleaning steps to prepare the joint. As copper sits exposed to the air, the surface starts to oxidize, preventing solder from bonding properly to the joint. Each pipe end and the inside of each fitting must be cleaned prior to soldering. A fitting brush can be used to clean the ends of the pipe. Another method is by using emery sanding paper. Sand the outside of a pipe by wrapping the paper around the pipe and rotating it around until shiny. It also can be rolled up with the grit exposed to slip into a fitting to clean the inside of the fitting. Soldering flux is then brushed on each joint before assembly to help further clean the copper and allow the solder to bond well.

36. Place the bottom pieces of copper pipe back into the wall through the isolator inserts.
37. Apply flux to the clean copper pipe and pipe fitting area to be soldered. If a day or more has gone by since cleaning, reclean the joint to remove new oxidation.
38. Assemble the pipe and the first 90° elbow together. Twist and push the fittings on the pipe to make sure they are fully inserted.

Name ______________________________

39. Assemble each of the joints with flux brushed on. At the top, slide the last horizontal piece of pipe through the stubout bracket at the top. Position each assembly so the vertical pipes are plumb inside the wall.

40. Place a fireproof shield or pad behind each joint to be soldered to protect the wood parts. This could be a fireproof cloth or simply a piece of aluminum coil stock.

41. Prepare to solder each joint by first unrolling about 6″ of the solder so the end can be placed on the joint without your hands getting too close to the flame. Always keep this much solder ready as each joint is soldered.

42. Next, the torch needs to be lit. Turn on the gas with the knob on the torch. With the cup of the striker close to the end of the torch, slide the arm of the striker across the inside of the cup to create a spark that ignites the flame.

43. Heat the pipe and fitting with the propane torch. Start with the flame on the pipe close to the fitting and then hold the flame on the fitting moving it around heating the whole fitting. Since heat rises, start at the bottom of the fitting to obtain even heating. The flux should bubble out of the joint indicating the cleaning action.

44. Check the temperature of the joint by touching the end of the solder to the top end of the fitting. If the solder doesn't melt, the joint is not hot enough. Continue to move the flame around the fitting, evenly heating it. When ready, the solder will melt and disappear into the joint. Keep the flame on the back of the fitting and the solder on the joint at the pipe. The solder will be drawn to the heat of the flame filling the whole joint with solder. See **Figure 7**.

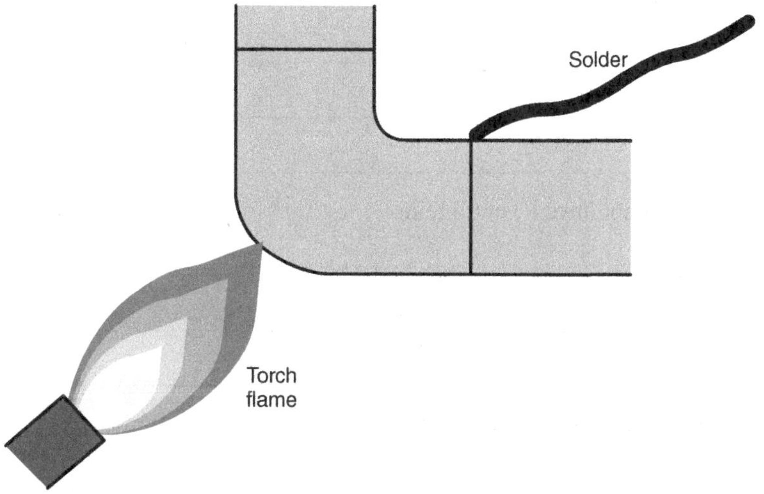

Goodheart-Willcox Publisher

Figure 7

> **NOTE**
> As a general rule for deciding the amount of solder needed for a good joint, the length of solder used should be equal to the size of the pipe being soldered. If soldering a 1/2″ pipe, each joint should use about a 1/2″ of solder.

45. While heating and preparing to solder, keep the flame on the pipe, not the solder.

46. After the joint is soldered and while it is still hot, wipe the joint with a damp cloth to clean any excess flux and solder. Be careful of hot metal.

47. Repeat these steps for each of the water lines soldering each joint.

48. Cap off the water lines going to the sink with copper end caps. Solder in place.

49. The final step is to solder the pipe to the stubout bracket. Check to make sure the desired length of pipe is out in front of the wall surface. Brush some flux around the pipe where it goes through the stubout. Just like the fitting, heat the pipe at the stubout and solder around the pipe bonding it to the stubout bracket.

50. Let the whole project cool and wipe off any excess flux residue for a clean, finished job.

51. Add a 2×4 piece of blocking behind the water lines as seen in **Figure 1**. Use two 16d nails on each end to fasten it to the studs.

52. Fasten the water lines to the blocking with two-hole copper pipe clamps with screws. See **Figure 8**.

Goodheart-Willcox Publisher

Figure 8

Review and Analysis

1. What would happen over time if there were no slope in the sink drain line downwards toward the main line?

2. What would happen if you did not have a vent line in your plumbing? What if your vent line became clogged?

3. What could happen if a PVC joint were not cemented properly?

4. What would happen if you did not clean all the oxidation off the copper pipe joint before soldering it?

Name ______________________________ Date ______________ Class __________

CHAPTER 7 Interior Insulation and Surfaces

Activity 7-1: Chapter 7 Review

Answer the following questions after studying the chapter.

Lesson 7.1

1. _____ Which person would be the best choice for a reference? (7.1-1)
   A. Parent
   B. Friend
   C. Previous employer
   D. Previous classmate

2. _____ *True or False?* Your cover letter should indicate when you are available to begin working at a new job. (7.1-1)

3. _____ When creating your cover letter, you should _____. (7.1-1)
   A. handwrite it on college-ruled paper
   B. proofread it for correct spelling and punctuation
   C. use a decorative font to highlight your creativity
   D. follow a standard form for personal letters

4. _____ On the day of the interview, it is important to _____. (7.1-1)
   A. be honest about your knowledge and abilities
   B. avoid making direct eye contact with the interviewer
   C. arrive exactly on time, neither early nor late
   D. refrain from asking questions about the workplace

Lesson 7.2

1. _____ Which of these conditions is an example of a musculoskeletal disorder? (7.2-1)
   A. Torn rotator cuff
   B. Migraine headache
   C. Type 2 diabetes
   D. Bronchitis

2. _____ Shivering, confusion, slurred speech, and drowsiness are symptoms of _____. (7.2-1)
   A. tendinitis
   B. heat exhaustion
   C. hypothermia
   D. frostbite

3. _____ *True or False?* In the winter, a dehydrated body maintains its temperature better than a hydrated body. (7.2-1)

4. _____ What is a symptom of dehydration? (7.2-1)
   A. Cold, clammy skin
   B. Itchy rash
   C. Blurred vision
   D. Impaired cognitive function

5. _____ While on a job in hot weather, it is recommended that you wear clothes that are _____. (7.2-1)
   A. light-colored and tight-fitting
   B. light-colored and loose-fitting
   C. dark-colored and tight-fitting
   D. dark-colored and loose-fitting

6. _____ *True or False?* Heat exhaustion is a more serious condition than heat stroke. (7.2-1)

Lesson 7.3

1. _____ *True or False?* An as-built drawing is produced after a phase of construction has been completed. (7.3-1)

2. _____ *True or False?* Studded basement walls must be insulated. (7.3-1)

3. _____ In a conditioned attic, _____ is often used between the rafters to act as an air barrier and a vapor retarder, keeping moisture from reaching a condensing surface. (7.3-1)
   A. rigid foam insulation
   B. loose insulation
   C. blanket insulation
   D. closed-cell spray foam insulation

4. _____ In new home construction, wood board paneling is commonly installed as a covering for _____. (7.3-2)
   A. ceilings
   B. walls
   C. floors
   D. ceilings, walls, and floors

Lesson 7.4

1. _____ *True or False?* Batt insulation and blanket insulation are made of the same material. (7.4-1)

2. _____ What is the most common wall and ceiling covering in residential construction? (7.4-2)
   A. Sheet vinyl
   B. Tile
   C. Drywall
   D. Wood planks

3. _____ *True or False?* When plank flooring is used, the planks are typically laid in uniform widths. (7.4-2)

4. _____ Which type of lumber defect is characterized by a variation from true or plane surface? (7.4-2)
   A. Knot
   B. Shake
   C. Wane
   D. Warp

5. _____ Cement board is _____. (7.4-2)
   A. highly flammable
   B. used for tile underlayment
   C. softened by exposure to water
   D. resistant to all adhesives

6. _____ Which of the following is a *true* statement about polyurethane adhesives? (7.4-3)
   A. They have many solvents.
   B. They can be applied at 95°F (35°C).
   C. They will bond with wood having up to 25% moisture content.
   D. They use a catalyst to activate the chemical that causes them to cure.

7. _____ *True or False?* Mastics are applied using a notched trowel. (7.4-3)

8. _____ Which of the following is a *true* statement about setting compounds? (7.4-3)
   A. They come in powder form and must be mixed with water.
   B. They shrink more than drying compounds as they harden.
   C. They have a very limited workability time of less than three minutes.
   D. They can be aggressively sanded after they have set.

9. _____ *True or False?* Lacquer is usually applied by brushing it onto wood. (7.4-3)

Lesson 7.5

1. _____ *True or False?* When using a utility knife to make a straight cut, after scoring one face of the sheet, the installer flexes the sheet in the direction of the scored face. (7.5-1)

2. _____ A drywall _____ is a small rotary power tool with a thin cutting bit to follow an object, allowing the cut to match the shape of the object. (7.5-1)
   A. power saw
   B. screw gun
   C. nipper
   D. router

3. _____ A finishing knife is used to _____. (7.5-1)
   A. shape pieces of drywall
   B. apply drywall compound
   C. smooth the edges of drywall
   D. remove dry compound before painting

4. _____ *True or False?* Short-nap roller covers are used for rough surfaces. (7.5-1)

5. _____ What is a power tool with a circular diamond blade that cuts tiles? (7.5-1)
   A. Scoring cutter
   B. Tile nipper
   C. Tile cutter
   D. Wet saw

Name ____

6. _____ Which tool is used to apply grout between tiles after they are installed? (7.5-1)
   A. Grout float
   B. Tile spacer
   C. Finishing trowel
   D. Notched trowel

7. _____ Which tool is used to install wood strip flooring? (7.5-1)
   A. Tapping block
   B. Flooring roller
   C. Flooring nailer
   D. Underlayment stapler

Lesson 7.6

1. _____ *True or False?* When moving heavy materials, OSHA recommends attaching handles or holders to loads whenever possible. (7.6-1)

2. _____ *True or False?* Fire prevention involves measures taken to minimize the damage caused by a fire that has already started. (7.6-2)

3. _____ In order for a fire to exist, there must be heat or an ignition source, fuel, and _____. (7.6-2)
   A. wood
   B. water
   C. oxygen
   D. electricity

4. _____ A fire involving rags soaked in oil and gasoline is considered a Class _____ fire. (7.6-2)
   A. A
   B. B
   C. C
   D. D

5. _____ What is the first step of the PASS technique? (7.6-2)
   A. Pinpoint the fire.
   B. Press the button.
   C. Point the extinguisher.
   D. Pull the pin out.

Lesson 7.7

1. _____ 10^3 = ___ (7.7-1)
   A. 3.33
   B. 100
   C. 300
   D. 1000

2. _____ What is the volume of a rectangular solid that is 7″ wide, 5″ tall, and 12″ deep? (7.7-1)
   A. 105 in^3
   B. 210 in^3
   C. 420 in^3
   D. 840 in^3

3. _____ If a triangular prism is 15″ long and 9″ high and has a base that is 7″ wide, its volume is _____. (7.7-1)
   A. 236 in^3
   B. 472.5 in^3
   C. 945 in^3
   D. 1890.5 in^3

4. _____ What is the volume of a cylinder that is 18″ long and has a face with a radius of 5″? (7.7-1)
   A. 282.74 in^3
   B. 450 in^3
   C. 900 in^3
   D. 1413.72 in^3

5. _____ *True or False?* The number 4^5 is spoken as "five to the fourth power." (7.7-1)

Lesson 7.8

1. _____ The detailed written documents that outline the specific requirements, materials, methods, and quality standards for construction are called the _____. (7.8-1)
   A. specifications
   B. general requirements
   C. as-built drawings
   D. scope of work

2. _____ *True or False?* The specifications are legally binding for the people responsible for carrying out the work. (7.8-1)

3. _____ The content under each heading in the specifications is divided into sections such as the scope of work, specifications of materials to be used, and _____. (7.8-2)
   A. landscaping
   B. basic information
   C. application methods and procedures
   D. finish carpentry and millwork

4. _____ *True or False?* The specifications may include information and requirements regarding building permits. (7.8-2)

Lesson 7.9

1. _____ On superinsulated homes, R-values in walls are often _____. (7.9-1)
   A. lower than R-20
   B. as high as R-30
   C. greater than R-40
   D. greater than R-60

2. _____ *True or False?* Deck-over roof insulation uses blanket insulation over the rafter sheathing. (7.9-1)

3. _____ *True or False?* Using best installation practices will yield a better return than simply adding more poorly installed insulation. (7.9-1)

4. _____ *True or False?* Robots are now being used in construction to screw sections of drywall into place. (7.9-2)

Name ______________________ Date ____________ Class ________

Activity 7-2: Skill Building—Scribing to Fit an Edge

Learning Outcomes

After completing this lab activity, you will be able to:

- Scribe a panel to fit an uneven surface.
- Use a scriber or compass to match an adjacent surface.

Introduction

While installing different materials next to each other in a corner, many times the surfaces are not smooth or perfectly in line. To make the transition look seamless and professional, the builder scribes one piece to fit the other. Scribing is marking a material to match the shape of an adjoining material so that when installed, they mate together in a corner. There are numerous ways to make this layout and cut accurate. In this activity, you will practice a couple methods commonly used to mark a matching cut for the two materials to fit skillfully together.

Instructor's Note: To help the students practice this skill, build a small corner with 2×4 framing to create an uneven surface. It can be placed on a work bench for easy access while doing the activity. **Figure 1** shows the example being used for the activity. Put drywall on one side of the corner with a 1/4″ shim behind the center of it to create an uneven surface to match. The shim can be moved up or down the corner to create different shapes to scribe to.

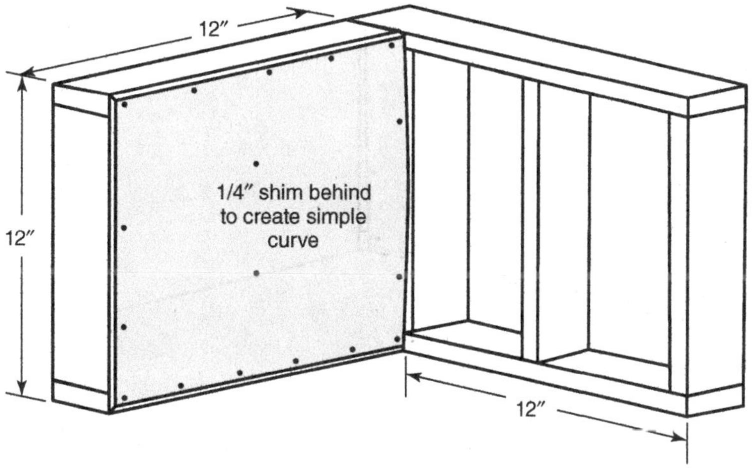

Goodheart-Willcox Publisher

Figure 1

Text Reference

Chapter 7, *Interior Insulation and Surfaces*

Equipment

The following tools and equipment are needed to perform these activities:

- 1×6×12″ pine
- 1 1/4″ drywall screws
- 1/4″ × 12″ × 12″ plywood
- Block plane
- Carpenter's pencil
- Compass or scribing tool
- Jigsaw
- Level, 2′
- Safety glasses
- Tape measure

Safety Review

Follow all safety procedures specified by your instructor and specified in any appropriate manufacturer's information. The following are some additional safety-related items to keep in mind:

- Wear safety glasses at all times.
- Be aware of the sharp point while handling the compass or scriber.
- Plan the cut with a jigsaw and make sure there is nothing under the piece near the blade.

Procedure

1. Place the piece of 1/4″ plywood on the right side of the corner and up against the drywall on the left. It should be touching the drywall in the center and there should be a gap open at the top and bottom of the corner. Use a screw to hold it in position to work on it.
2. For the first method of scribing, hold a small block of wood that is 3/4″ thick at the top of the corner against the drywall. A small block like this works well for when you need to scribe to a smooth surface. Place the pencil against the side of the piece of wood, touching the plywood, **Figure 2**. Slide the piece of wood and pencil down the corner, marking the shape of the drywall on to the plywood.

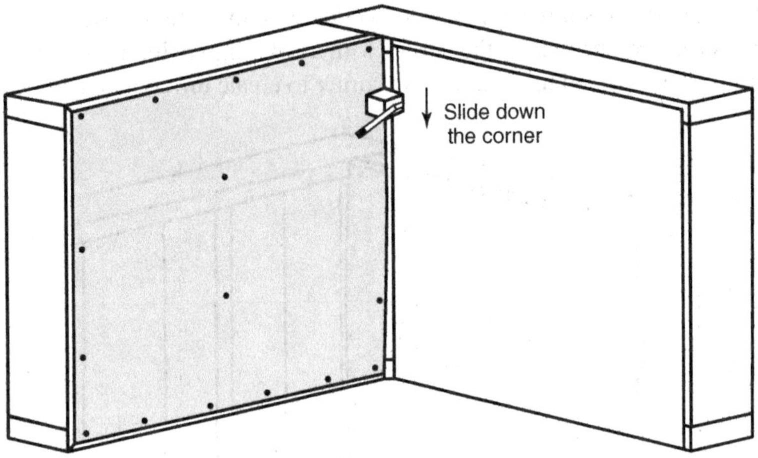

Goodheart-Willcox Publisher

Figure 2

3. Using the block plane, shave the edge of the plywood until it follows the shape of the pencil mark scribed from the drywall. Slide the plywood into the corner to check the fit. If there are still small gaps, shave the high spots off for a uniform fit. Then the plywood panel is ready to install.

Scribing to Specific Dimensions

Next, you will scribe a piece to fit a specific dimension, which happens when installing a cabinet or wood wainscot. Use the 1×6×12″ piece of pine for this. The final placement of the board for installation will be 5″ from the drywall surface at the top edge. The scribe will accomplish this final placement.

4. Measure and place a mark 6″ from the drywall face at the top of the wall. If the pine board is held at the 6″ mark and scribed with the compass set at 1″, it will be right at the 5″ desired distance from the corner after scribing and cutting.
5. Place the 1×6 board vertically in the corner, lining up the right edge of the board with the 6″ mark. Use the 2′ level to ensure the board is plumb. Hold the board in place with a screw at the top and bottom while marking.

6. Set the compass or scriber with a tape measure so the distance from the needle to the pencil point is 1″, **Figure 3**. A compass or scriber is the best tool to scribe to an uneven surface or when you need to reach a specific dimension.

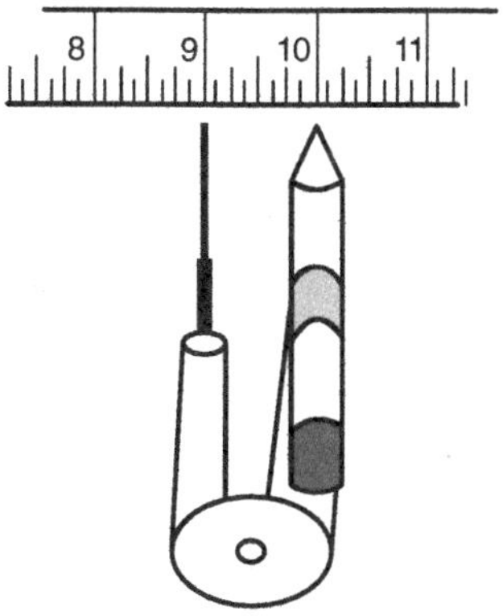

Goodheart-Willcox Publisher

Figure 3

7. At the top of the corner, place the needle point against the drywall and the pencil out onto the board in a level position. Slide the compass down the corner, keeping the needle point against the drywall and the pencil making a mark on the board. Keep the needle point and the pencil even and level with each other as you move it down the corner to get the best match of the shape being scribed, **Figure 4**.

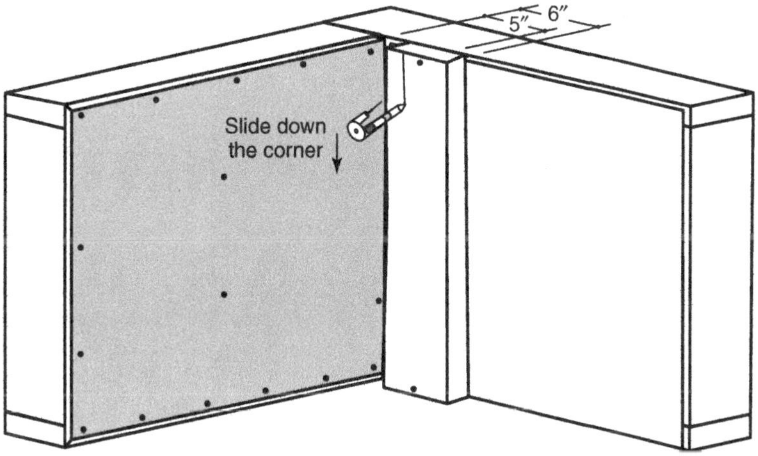

Goodheart-Willcox Publisher

Figure 4

8. Take the pine off the wall and use the jigsaw to cut along the scribe mark to create the matching shape of the wall. Use the block plane to smooth out the edge for a tight fit. Put the board back in place to check the fit and if it is in line with the 5″ mark as desired.

Review and Analysis

1. Describe two situations when a builder would scribe a corner for a good fit.

2. Why is it important to keep the compass level across the points when making the scribe?

3. When does the block method of scribing work the best?

Activity 7-3: Skill Building—Cutting Drywall

Learning Outcomes

After completing this lab activity, you will be able to:

- Cut drywall using different methods during installation procedures.

Introduction

Drywall is the most common interior surface finish used in construction. On many projects, a carpenter will be required to install or repair drywall surfaces. Understanding how to cut drywall while hanging it is an important skill. This activity will provide practice using several methods used to cut drywall to fit a space or to create openings needed for electrical boxes or other fixtures.

Text Reference

Chapter 7, *Interior Insulation and Surfaces*

Equipment

The following tools and equipment are needed to perform these activities:

- Carpenter's pencil
- Compass
- Drywall jab saw
- Drywall router
- Drywall sheets, 4×8
- Drywall T-square
- Safety glasses
- Tape measure
- Utility knife
- Workbench

Safety Review

Follow all safety procedures specified by your instructor and specified in any appropriate manufacturer's information. The following are some additional safety-related items to keep in mind:

- Wear safety glasses at all times.
- Be sure to retract the utility knife after each use.
- Always cut away from you when using the utility knife.
- Wear a dust mask when using the drywall router.

Procedure

For this activity, use the walls of the corner unit being built to practice cutting and hanging drywall. Have at least one electrical box installed in the wall to practice the use of the drywall router.

Cutting Drywall with a Utility Knife

1. To cut the first piece from the drywall sheet, first place the sheet upright leaning against a wall for easy access. On the front of the sheet, measure 12″ from one end and hold the T-square so the blade of the square goes across the whole sheet in line with the measurement.
2. While holding the square tight against the sheet, cut along the blade of the square into the face of the drywall using the utility knife. The cut does not have to be too deep into the sheet but must go through the paper face.

3. Pull the sheet away from the wall and bend the sheet backward, opening the cut. The drywall should snap at the cut line. Bend the piece back at about a 90° angle and cut the backing paper with the utility knife. Use the sheet as a guide to slide the knife along the bend to make the cut. There should be a piece of drywall now 12″ wide by 48″ long.
4. On the workbench, cut the drywall piece in half for two pieces measuring 12″ × 24″ each. Mark the 24″ length to be cut and hold the square across the piece to guide the cut. After making the cut, flip the piece over and bend it at the cut to snap it and cut the backer paper with the knife to separate the pieces.
5. Using one piece that was just cut, layout a center point 6″ from the end of the piece and 6″ from the side. Use the compass to draw a 6″ circle on the face, **Figure 1**.

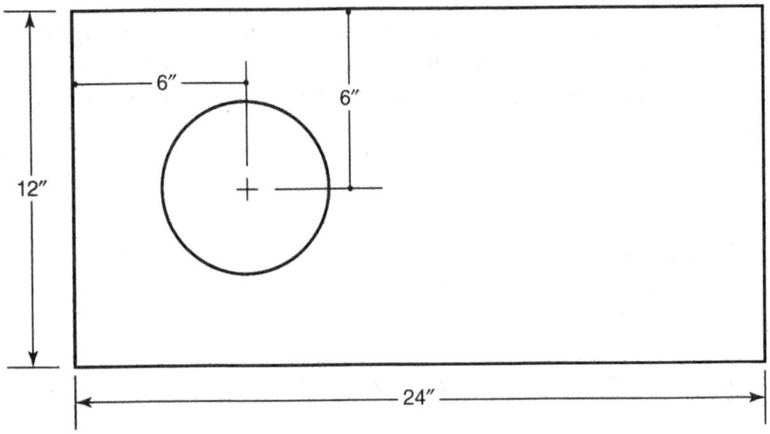

Goodheart-Willcox Publisher

Figure 1

Cutting Drywall with a Jab Saw

1. A drywall jab saw is one of the most used tools to cut openings for fixtures in a wall or ceiling during installation. While holding the piece up at an angle on the bench top, take the jab saw and poke it through the drywall at a point on the circle. Without too much pressure, push and pull the saw to cut out the circle. The harder the saw is pushed, the harder it is to cut with it. Just let the saw do the cutting as it slides in and out of the cut around the circle.
2. When the cut is complete and the circle falls out, there may be some rough spots in the paper on the back side of the cut. Take the jab saw and hold it against the back side of the cut at an angle with the handle at the center of the circle. Draw it toward the front of the piece, making sure to drag the teeth along the back edge of the cut. This will clean up the ragged edges on the back of the cut.
3. On the other end of the piece, layout a 4″ square centered in the width and 4″ from the end, **Figure 2**.

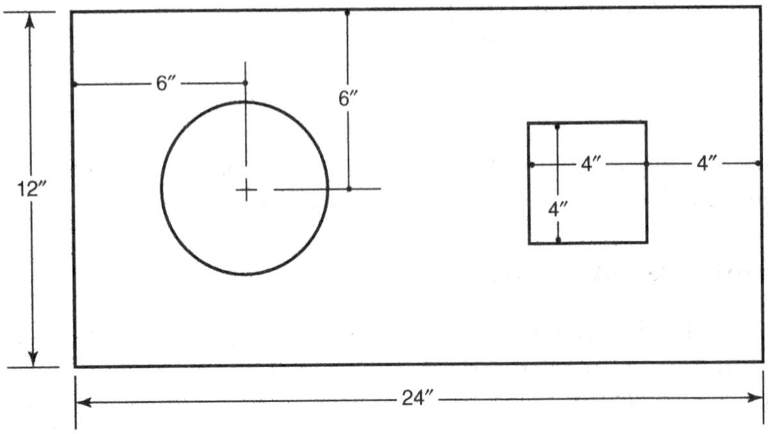

Goodheart-Willcox Publisher

Figure 2

Name ______________________________

Making Internal Cuts with a Utility Knife

1. A jab saw may not always be available on a job, so a carpenter should be familiar with making an internal cut without it. Use the utility knife to make a cut around the perimeter of the square, carefully starting and stopping at each corner. Then, make a cut across the square at both diagonals from corner to corner. Go over these cuts a couple of times each so the knife is about 1/4″ deep.
2. With the back end of the knife or a hammer, lightly hit the center of the square and break out each of the four triangles formed by the cuts. Then, use the knife to cut the backer paper to remove each of the triangles. Use the knife to clean up the perimeter of the cutout making sure the edges of the drywall are square in the cut.
3. Many times, a small strip must be removed from a piece of drywall to make it fits the space being filled. With a little practice, this can be done quickly and easily with a tape measure and utility knife. To do this, first fasten or clamp the piece of drywall on the bench with 3″ hanging over the edge of the bench. Hold the tape measure in one hand with the index finger on the edge of the piece. Pull the tape out so the 2 1/2″ mark is at the edge of the piece and perpendicular to the edge and lock it. Using your other hand, hold the knife against the end of the tape measure, **Figure 3**. With both hands acting together, make a cut along the edge of the drywall piece as the index finger of the hand holding the tape body slides along the edge and the knife stays at the end of the tape making the cut.

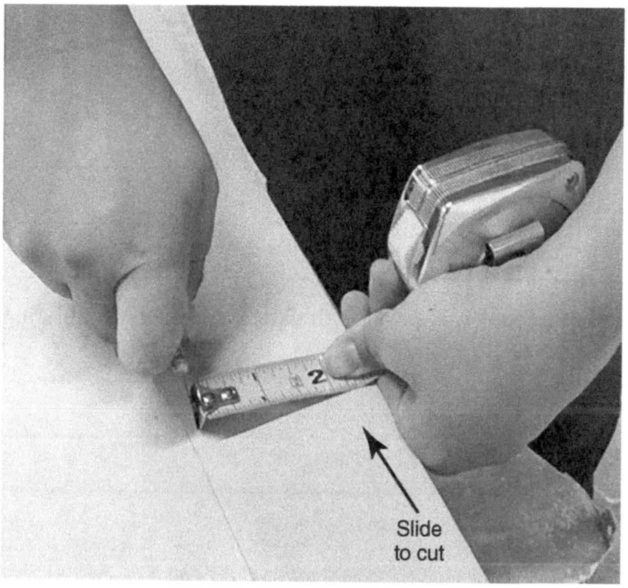

Goodheart-Willcox Publisher

Figure 3

4. A skillful drywall installer will be able to use this method to cut a strip at just about any width. After the cut is made to the full length of the piece, snap it and use the knife to cut the backer paper behind the piece.

Cutting Drywall with a Router

1. The last type of tool an installer may use to cut drywall is the drywall router, which is one of the most efficient methods of cutting drywall.
2. On the corner unit, measure where the center of the electrical box is from the edge and bottom of the sheet to be installed. Mark that on the piece of drywall before putting it on the wall.
3. Be sure the electrical box wires are pushed into the box as far as possible to ensure the drywall router bit does not damage them.
4. Put the piece of drywall in place and fasten it only along the top of the sheet. It should be hanging down the wall and laying on the face of the electrical box.
5. Set the depth of the drywall router so the bit will cut 1/4″ deeper than the drywall thickness. If the drywall is 1/2″ then add 1/4″ to that and set the depth of cut at 3/4″. The bit will go through the drywall and ride on the box to make the cut. Be sure to put on the dust mask before making the cut.

6. With the drywall router running, push the bit through at the center mark of the box on the drywall face. Move the router to the side until it hits the inside edge of the box. Pull the router out of the wall slowly until the bit goes past the box edge. It is now on the outside of the box. Push the router back into the wall so the end of the bit can guide the rest of the cut.
7. Move the router counterclockwise around the box so the bit automatically helps keep itself on the edge of the box. Go slow enough to feel when the bit comes to a corner and goes around it. Always keep enough pressure on the router to keep it against the box as it is guided through a cut. When the cut is complete around the box, the cutout piece should fall out and the drywall will go back against the wall studs. The cut should be guided by the sensation of the router riding along the outside of the box since the guiding action of the bit is behind the piece and cannot be watched to control it. Now, the drywall can be fastened the rest of the way onto the wall framing.

Review and Analysis

1. Describe how deep a cut needs to be made to ensure drywall breaks off easily.

____

____

____

2. When cutting into drywall, what other building components besides electrical wires should a carpenter be aware of?

____

____

____

3. Why is the bit on the drywall router set 1/4″ past the thickness of the drywall being cut?

____

____

____

Name ______________________ Date ______________ Class ______________

CHAPTER 8 Interior Finishes

Activity 8-1: Chapter 8 Review

Answer the following questions after studying the chapter.

Lesson 8.1

1. _____ Employers typically work with health insurance companies to provide coverage for a fee called the insurance _____. (8.1-1)
   A. benefit
   B. contribution
   C. coverage
   D. premium

2. _____ Which method of employer contribution to a retirement plan involves the employer contributing the same amount to the plan that the employee contributes? (8.1-1)
   A. Employer match
   B. Profit sharing
   C. Fixed percentage
   D. Variable percentage

3. _____ *True or False?* If an employer offers retirement plans, they must make contributions to each employee's account on a regular basis. (8.1-1)

4. _____ If a worker is typically paid $30 per hour during a 40-hour work week, the employer must pay them at least _____ per hour when they work more than 40 hours in a work week. (8.1-1)
   A. $15
   B. $30
   C. $45
   D. $60

5. _____ *True or False?* Federal law mandates that company vehicles can only be used on a jobsite. (8.1-1)

Lesson 8.2

1. _____ *True or False?* Collaboration often relies on mutual respect for achieving the overall outcome. (8.2-1)

2. _____ *True or False?* Keeping projects on schedule often requires a give-and-take relationship across trades. (8.2-1)

3. _____ *True or False?* Although being a team player can help move a job along smoothly, it will limit your future professional opportunities. (8.2-1)

4. _____ Exhibiting responsibility on the job will _____. (8.2-1)
   A. make it difficult to allocate resources effectively
   B. build trust in professional relationships
   C. interfere with the work of other professionals
   D. reduce the size of your professional circle

Lesson 8.3

1. _____ What is an example of casework? (8.3-1)
   A. Bathroom cabinet
   B. Baseboard trim
   C. Chair rail
   D. Stair railing

2. _____ *True or False?* Stairs are most commonly installed by a carpenter's assistant. (8.3-1)

3. _____ When working with stairs, what is the overall vertical distance from one floor to the next called? (8.3-1)
A. Total rise
B. Total run
C. Unit rise
D. Unit run

4. _____ The riser height is the _____. (8.3-1)
A. total rise
B. total run
C. unit rise
D. unit run

5. _____ *True or False?* In residential structures, treads are usually no more than 12″, excluding nosing. (8.3-1)

6. _____ Which type of stairway includes one or more landings where the direction of the stair run is usually changed? (8.3-1)
A. Geometrical stairway
B. Platform stairway
C. Straight run stairway
D. Winding stairway

7. _____ The molding that covers the joint where the wall intersects the finish flooring is the _____. (8.3-1)
A. baseboard
B. base shoe
C. chair rail
D. crown molding

8. _____ *True or False?* Most of the cabinetwork for residential buildings is constructed at the jobsite. (8.3-1)

9. _____ When working with kitchen cabinets, the vertical distance between the top of the base unit and the bottom of the wall unit is normally _____. (8.3-1)
A. 12″
B. 14″
C. 18″
D. 24″

10. _____ A wall cabinet installed above a refrigerator should typically be installed _____ above the finished floor. (8.3-1)
A. 62″
B. 68″
C. 72″
D. 84″

Lesson 8.4

1. _____ FAS lumber must be at least _____. (8.4-1)
A. 4″ wide and 6′ long
B. 4″ wide and 8′ long
C. 6″ wide and 6′ long
D. 6″ wide and 8′ long

2. _____ *True or False?* Lumber in the No. 1 common grade is expected to yield 66.66% clear cutting. (8.4-1)

3. _____ *True or False?* A flush door is made with either a solid or hollow core. (8.4-1)

4. _____ It is common practice for closet doors to be _____ wide. (8.4-1)
A. 2′-0″
B. 2′-6″
C. 2′-8″
D. 3′-0″

5. _____ Which type of molding is used as a base cap to dress off the top of the base when the base's top edge is squared off? (8.4-1)
A. Quarter round
B. Half round
C. Three-quarter round
D. Full round

6. _____ Which statement about solid-surface countertops is true? (8.4-1)
A. They are porous and stain easily.
B. They are made of solid wood.
C. They can be made to look like stone.
D. They must be screwed into the base cabinet(s).

7. _____ *True or False?* Cabinet screws are designed to be used without first predrilling pilot holes. (8.4-2)

8. _____ Which statement about molly bolts is true? (8.4-2)
A. They are a type of light-duty hollow wall anchor.
B. As they are tightened, their sleeve flares against the back of the wallboard.
C. They use a plastic sleeve as a compression device to distribute the load.
D. They are also known as toggle bolts.

Name ____

Lesson 8.5

1. _____ What type of saw has 8 teeth per inch? (8.5-1)
   A. Backsaw
   B. Coping saw
   C. Finishing saw
   D. General use crosscut saw

2. _____ Which type of saw has a thin, flexible blade that is pulled tight by the saw frame and has a blade that has 15 teeth per inch? (8.5-1)
   A. Backsaw
   B. Coping saw
   C. General use crosscut saw
   D. Pull saw

3. _____ Which type of plane is usually selected for general-purpose work? (8.5-1)
   A. Block plane
   B. Fore and jointer plane
   C. Jack plane
   D. Smooth plane

4. _____ *True or False?* The try square is the most basic version of a marking square. (8.5-1)

5. _____ *True or False?* The T-bevel has an adjustable blade. (8.5-1)

6. _____ Which statement about the saber saw is true? (8.5-1)
   A. It is specifically designed for heavy-duty work.
   B. The stroke of its blade is about 1″.
   C. It operates at a speed of approximately 2500 strokes per minute.
   D. Its blade cuts on the downstroke.

7. _____ *True or False?* Belt sanders are used to lightly sand small surfaces. (8.5-1)

Lesson 8.6

1. _____ *True or False?* An acute health condition is characterized by a health issue that persists over a long period of time. (8.6-1)

2. _____ Volatile organic compounds (VOCs) contain which element? (8.6-1)
   A. Carbon
   B. Lead
   C. Oxygen
   D. Sodium

3. _____ *True or False?* Painting materials are a major source of VOCs. (8.6-1)

4. _____ Prolonged exposure to silica dust can cause diseases of the _____. (8.6-1)
   A. heart
   B. liver
   C. lungs
   D. pancreas

5. _____ *True or False?* Simply using PPE correctly will fully protect you from all hazards on the job. (8.6-1)

Lesson 8.7

1. _____ If six windows have areas of 9.4 sq ft, 12.5 sq ft, 12.5 sq ft, 12.5 sq ft, 14.3 sq ft, and 15.2 sq ft, what is the average area of this group of windows? (8.7-1)
   A. 12.5 sq ft
   B. 12.7 sq ft
   C. 13.8 sq ft
   D. 76.4 sq ft

2. _____ *True or False?* When determining an average, the quotient should always be rounded off to the nearest tenth. (8.7-1)

3. _____ 0.42 = _____ (8.7-1)
   A. 0.42%
   B. 4.2%
   C. 42%
   D. 420%

4. _____ 79% = _____ (8.7-1)
   A. 0.79
   B. 7.9
   C. 79
   D. 790

5. _____ What is 13% of $528.47? (8.7-1)
A. $2.46
B. $40.65
C. $68.70
D. $79.87

Lesson 8.8

1. _____ Where should you look to find information about the edge profiles that should be used for countertops on a job? (8.8-1)
A. Finish schedule
B. Floor plan
C. Interior elevations
D. Specifications

2. _____ *True or False?* A finish schedule is typically a table or chart that lists the finishes for each room or area. (8.8-1)

3. _____ For information about cabinetry layouts, such as dimensions, locations, and styles, consult the _____. (8.8-1)
A. finish schedule
B. floor plan
C. interior elevations
D. specifications

Lesson 8.9

1. _____ *True or False?* Source reduction involves dealing with waste that has already been created. (8.9-1)

2. _____ Deconstruction involves _____. (8.9-1)
A. completely demolishing every part of a building
B. dismantling buildings to salvage usable components
C. dismantling buildings and disposing of all components
D. leaving buildings intact so that they can be renovated

3. _____ *True or False?* Recycling involves processing materials for reuse. (8.9-1)

Name ______________________ Date ______________ Class __________

Activity 8-2: Skill Building—Installing Trim

Learning Outcomes

After completing this lab activity, you will be able to:

- Prepare the basic joints required to install finish trim.

Introduction

The final stages of completing a house include installing the trim and woodwork to cover transitions and corners. This work requires a skilled carpenter to achieve the desired appearance. Care must be taken when cutting and fitting the trim pieces to obtain the best appearance. This activity will provide some practice toward working with woodwork and trim.

Instructor Note: To prepare for this exercise, build a box of 2×6 lumber on a plywood base to create outside corners and inside corners to fit the trim on for this practice, **Figure 1**. The edges of the 2×6 lumber should be jointed to make accurate square corners. Students can work with short pieces on one corner or fill in around the box.

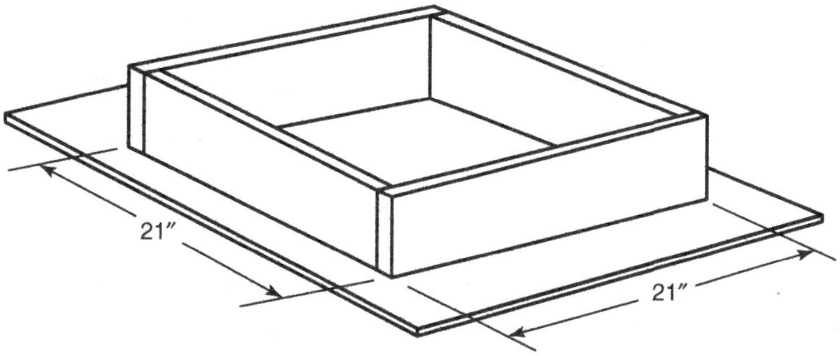

Goodheart-Willcox Publisher

Figure 1

Text Reference

Chapter 8, *Interior Finishes*

Equipment

The following tools and equipment are needed to perform these activities:

- 4d finish nails
- Carpenter's pencil
- Coping saw
- Cove molding
- Hammer
- Nail set
- Power miter saw
- Ranch baseboard
- Safety glasses

Safety Review

Follow all safety procedures specified by your instructor and specified in any appropriate manufacturer's information. The following are some additional safety-related items to keep in mind:

- Wear safety glasses at all times.
- Be sure your hands are not near the blade when cutting on the miter saw.
- Never cross your arms in front of the miter saw to hold the pieces being cut.

Procedure

Mitered Corner

1. Start with a piece of baseboard a couple inches longer than the length of one side of the box. Set the miter saw at 45° to the right side of the saw and lock it in place. Stand the baseboard up against the fence of the saw to the left of the blade so the cut will only remove enough length to produce a full miter on one end. Make the cut slowly while keeping the baseboard tight against the fence.
2. Set the baseboard piece on the box at one corner and line up the inside corner of the miter cut with the corner of the box. While holding it there, mark the other end of the baseboard piece right at the corner of the box, **Figure 2**. Put a light mark from the corner out and away roughly at the angle of the miter for that corner. This will remind you which way the miter has to be made.

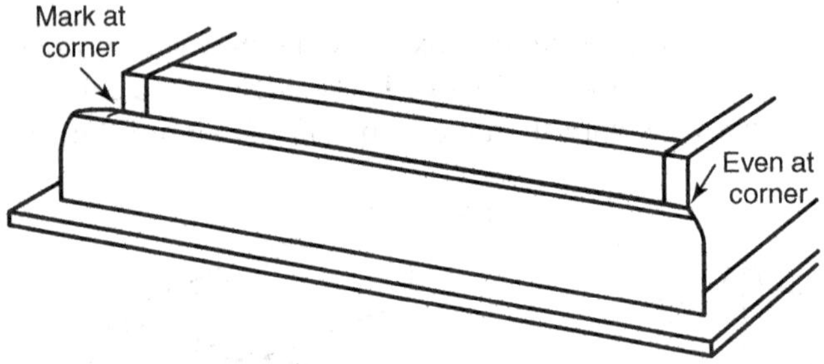

Goodheart-Willcox Publisher

Figure 2

3. Take the piece back to the miter saw. This cut will be done in the opposite direction from the first miter. Unlock the miter saw and swing it to 45° on the left side of the saw. Put the baseboard on the fence as before but on the right side. Without holding the trigger of the saw, pull the saw down so the blade almost touches the top edge of the baseboard. Slide the baseboard into position, lining up the mark with the side of a tooth on the blade. Be sure to cut on the waste side of the mark.
4. Carefully line up the baseboard on the box with the inside corner of each miter right at the corners of the box. Nail it in place with the 4d finish nails. Place a nail about an inch back from the corner near the top and a second nail just below the first near the bottom of the baseboard. Do the same on the other end of the piece to secure the whole piece. Hammer the nail in until just the head of the nail is left out of the piece. This is finish material, so care must be taken not to hit the baseboard with the hammer and put dents in it.
5. Use the nail set to drive the nail the rest of the way into the baseboard. The head of the nail should be just below the surface. The little hole created will be filled with a paste wood filler or painter's caulk for a smooth finished surface.
6. Prepare a miter cut on the end of another piece of baseboard just like the first. Hold this piece on the box with the miter held tight against the miter of the first piece. Check the fit and note where the miters touch. If the miter cuts fit tight along the whole width of the cut, it is a good fit and the next step can be skipped.
7. If the miter cuts do not align smoothly, take the piece back to the miter saw to re-cut the miter. If the miters touched at the back of the cut near the box corner, move the blade slightly back, away from the miter and make another cut taking off only a little length of the piece. If the miters touched at the front of the cut near the point, move the blade slightly forward toward the miter and make another cut. The corner should fit well now.
8. Mark the second piece for length now by holding the piece in place against the first piece at the box and mark the other end as in Step 2. Swing the miter saw to the other side at 45° and cut the piece while carefully lining up the saw blade with the mark.
9. Nail this piece in place with the finish nails and set the nails as in Steps 4–5. Continue around the box, repeating each step until the baseboard meets the first piece.
10. If there is a small gap at the point of the miter, roll the nail set up and down the corner to push the wood fibers together and close the gap. The joint will look much more accurate after pushing it together like this.

Name ___________________________________

Scarf Joint

11. When there is a long wall, it may take more than one piece to cover the whole distance. A joint needs to be made bringing two or more pieces together to create the length desired. Two pieces are not just butted together because changes in humidity will cause the joint to open and close throughout the year. Instead, a scarf joint with a 45° angled cut to overlap each piece is used so the joint won't open and be visible if the wood shrinks.
12. Carefully remove one of the baseboard pieces from the box to practice this joint. Be sure to pull all the nails from the piece or the box.
13. Set the miter saw to a 45° angle to the right side. Measure and mark the piece at 12″ from the right end. Cut the piece and fasten it to the box with the miter cut in the center of the side of the box as seen in **Figure 3**.

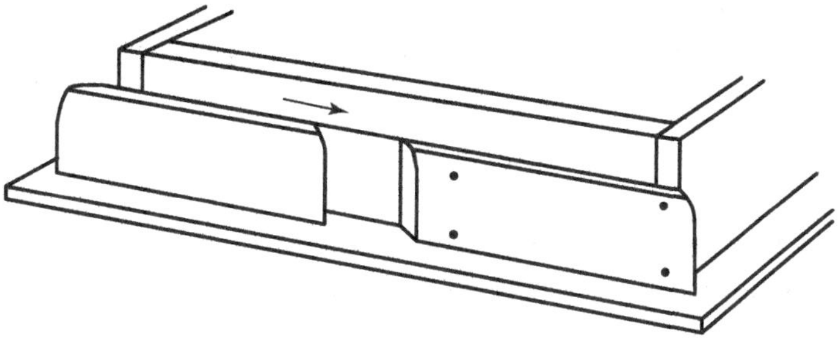

Goodheart-Willcox Publisher

Figure 3

14. Use the remaining piece to make the matching cut for the scarf joint. Adjust the saw toward the fence by about 1°. This will adjust the cut so the front of the miters touch completely and give a tight joint. Cut just a small amount off the length.
15. Put the piece on the box, overlapping the other one at the miters. Line up the faces of the two pieces for a smooth transition across the joint. Nail it in place close to the joint to form a tight joint.

NOTE

When placing scarf joints in trim in a room, the overlap should always be done away from the primary sight line when entering the room. For the best appearance, you should be looking across the miter and not into it when cut correctly.

Coping a Joint

16. When installing trim in an inside corner, the coped joint is made so the pieces overlap each other and keep the joint looking tight when humidity changes occur. A plain miter will open and close with humidity changes and be visible.
17. Use a piece of cove molding for this joint to practice coping the joint. Start by cutting a piece to fit inside the box across one side. Use a square cut on each end to fit into the corners, **Figure 4**. Place it in the bottom of the box and tack it in place with a couple of finish nails.

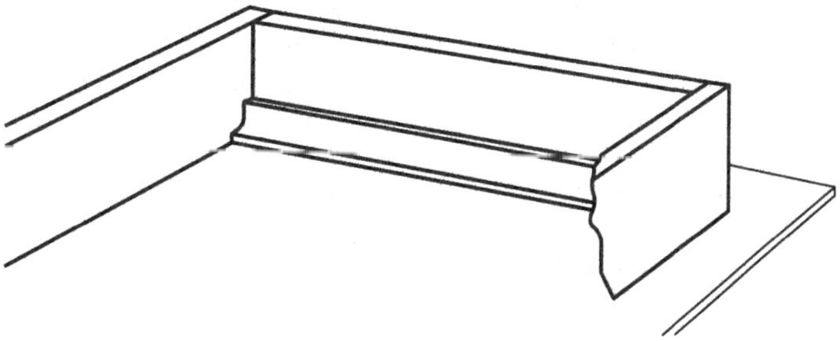

Goodheart-Willcox Publisher

Figure 4

18. Cut another piece to fit an adjacent side of the box. Set the saw to a 45° angle to the left side of the saw. Holding the molding on the left side of the blade, make a miter cut on the end. The shape that is visible along the cut is the line that will be cut while coping the joint. **Figure 5** shows what to follow to make the cut.

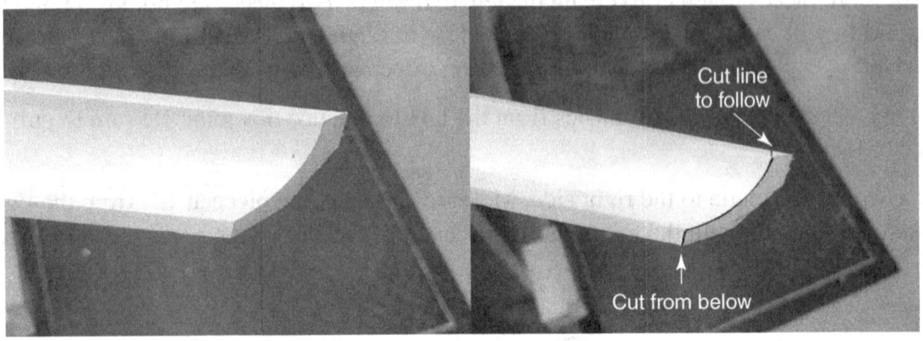

Goodheart-Willcox Publisher

Figure 5

19. Hold the molding in a comfortable position on a bench or a sawhorse directly in front of you. Hold the coping saw at the top of the trim, **Figure 6**. The blade should be held level across the trim from front to back and square with the front edge to start the cut.

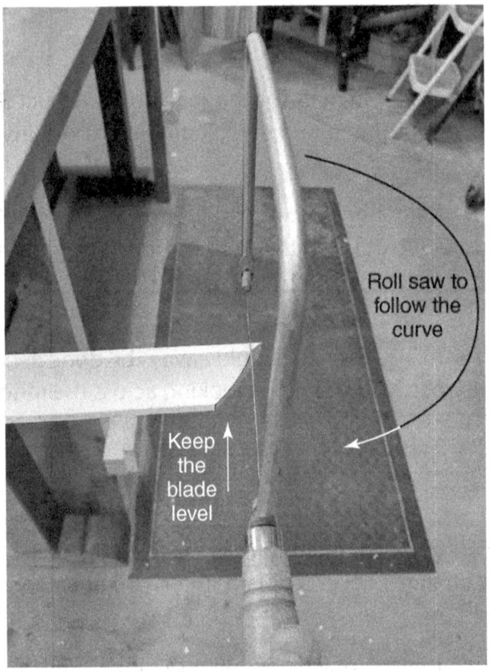

Goodheart-Willcox Publisher

Figure 6

20. Start making the cut with light pressure down into the piece. Do not force the saw. Let the saw do the cutting. Slowly roll the saw down and to the right as the cut progresses, following the profile on the cut. Keep the saw blade level as it follows the curve. Cut until the saw reaches the vertical line of the profile. Stop and back the blade out of the cut.

21. Turn the saw over to cut from the bottom of the piece. Line up the blade with the vertical line of the profile and cut straight up to the end of the curve cut just completed. Again, keep the saw level and square with the face of the molding. When the curved cut is reached, the piece being cut should fall out.

Name ______________________________

22. This coped piece should fit into the other piece in the box and create a tight joint, **Figure 7**. If it doesn't close tight, some additional cutting may be needed under the curve to get it to sit down tight against the first piece. This adjustment can be made with a utility knife or a rasp to remove a little more material for a tight joint.

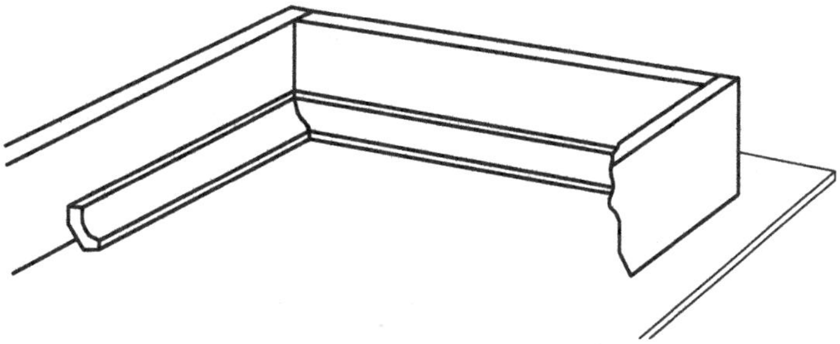

Goodheart-Willcox Publisher

Figure 7

23. Try cutting another one to practice how the coping saw should be held to get the correct curved cut. When comfortable making that cut, make a piece to fit from corner to corner in the box with the coped joint on one end and a square cut on the other to fit into the corner.

Review and Analysis

1. Why is the head of the finish nail driven below the surface of the trim?

____

____

____

2. How does the direction of the lap in the scarf joint affect the final appearance of the trim?

____

____

____

3. Describe why the coped joint is better than a miter joint in an inside corner.

____

____

____

Notes

Name ______________________ Date ____________ Class ________

Activity 8-3: Skill Building—Developing a Finish Schedule for a Room

Learning Outcomes

After completing this lab activity, you will be able to:

- Create an interior room finish schedule.
- Specify a building product to be used for a project.

Introduction

During the design process, an architect not only creates the layout of the rooms in a building and how they all fit together but may also help the customer choose finishes for the rooms. Everything from flooring, wall finishes, and the woodwork will be chosen and specified. To put all this information in an easy-to-read format, a finish schedule is created as part of the set of prints. This schedule is a table that organizes the information for each surface being finished. Many times, a contractor will follow the same process for a customer even if an architect is not involved. In this activity, you will create a finish schedule for your rooms drawn in **Activity 3-2** to practice the process.

Text Reference

Chapter 8, *Interior Finishes*

Equipment

The following tools and equipment are needed to perform these activities:

- Paper
- Pencil

Procedure

1. When an architect creates a set of plans for a home, the compass directions are used to help make instructions about the building and rooms easy to locate. When the building walls do not line up square with the north-south or east-west directions, a plan north is chosen. Typically, the wall that lines up closest with the north direction will be labeled the *North Wall* for ease of labeling. This side will determine the plan north which may be offset from the true north direction on a compass. Decide which will be the plan north of your rooms to help keep labels organized.
2. Use **Figure 1** to organize the information prepared to define all the finishes in your rooms.

Room Finish Schedule									
Room	North Wall	East Wall	South Wall	West Wall	Floor	Ceiling	Opening Trim	Baseboard Trim	Ceiling Trim
Example Room	Paint Valspar Eggshell	Paint Valspar Eggshell	Paint Valspar Eggshell	Paint Valspar Eggshell	Hardwood, Oak Prefinished	Paint Valspar Flat White	Ranch Casing 2 1/4"	Ranch 3 1/2"	Cove Molding 1 3/8"

Goodheart-Willcox Publisher

Figure 1

Name ______________________________

3. Look over the rooms that were drawn for **Activity 3-2** and identify the different finishes. Begin with what has been used to finish the walls. Is there just paint, or is there wallpaper or other woodwork like a wainscot? A wainscot is a wall finish that only covers the lower portion of a wall and is capped with a chair rail or similar trim.
4. Start with your plan *North Wall* and fill in the information needed on the table. Do the same for each wall around the room. Sometimes one or two walls may have different finishes than the others for a design accent.
5. Determine your flooring materials. How is the floor covered? Hardwood floors or carpet are very common for bedrooms. Tile or sheet flooring might be found in bathrooms. Determine the flooring material and record the information in the table.
6. What is on the ceiling? In older homes there may be a tiled ceiling or even a suspended ceiling. In most newer homes the ceiling is painted drywall. Determine the ceiling materials and record the information in the table.
7. To finish the schedule, look at the trim used to finish around the openings and the floor. Doors and windows will have trim around them, spanning the space between the jambs and the wall surfaces. Around the floor will be baseboard to finish that corner and there may be trim around the ceiling to add to the decoration. Research trim profiles and identify the terms that name the materials being used for each location.
8. For a little challenge, carefully look over the color of the walls. Go to a local lumber yard or paint store and look through the paint chips. Paint sellers will have a large rack of samples with all shades of colors to help customers choose; these are referred to as paint chips. Choose four of them you think are closest to the wall color at home. When you return home, see how well you matched the color from memory. Did you get an exact match?
9. To complete the activity, research three items from the finishes listed and create a specification for each. A specification will define the details of a product to use in the room so the contractor and owner are in agreement when the job starts. Choose a flooring type used in one of your rooms, use a wall finish, and specify the trim being used for the openings. Research each product chosen and create the following list of information:
   - Trade name of the item.
   - Manufacturer and model number.
   - Dimensions of the item if applicable.
   - Color from manufacturer's available choices.
   - Surface finish—gloss, semi-gloss, eggshell, or matte are names used.
   - For wood trim, define how it will be finished—painted, stained, etc.
   - Any other information that helps to identify the product.

Review and Analysis

1. How does a finish schedule help ensure customer satisfaction on a job?

____

____

____

2. Were you surprised by the variety of options for finishing products?

____

____

____

3. If you were to redesign your room, what products would you change?

____

____

____

Notes

Name ______________________ Date ______________ Class ____________

Activity 8-4: Skill Building—Building a Stair

Learning Outcomes

After completing this lab activity, you will be able to:

- Layout a stair stringer.
- Fabricate a jobsite stair.

Introduction

Stair construction is done by the most skilled carpenters. Not only do they need to know the construction process but also the design process and the calculations involved. Before any material is cut for this activity, careful measurements and calculations must made to determine the design of the stair. This activity will get you started on building those skills with the construction of a basic stair layout.

Text Reference

Chapter 8, *Interior Finishes*

Equipment

The following tools and equipment are needed to perform these activities:

- 2×10×8′ stringer material
- 2×4×30″ lumber
- Carpenter's pencil
- Circular saw
- Framing screws, sized 2 1/2″ and 1 5/8″
- Framing square
- Handsaw
- Level, 2′
- OSB tread material, 11″ × 48″ × 3/4″
- Safety glasses
- Stair gauges

Safety Review

Follow all safety procedures specified by your instructor and specified in any appropriate manufacturer's information. The following are some additional safety-related items to keep in mind:

- Wear safety glasses at all times.
- Keep hands clear of the circular saw blade when making cuts.
- Ensure the blade guard on the circular saw is working properly.
- When working with the handsaw, be careful of the teeth when guiding the cut.

Procedure

Calculations

The first measurement needed for a stair design is the total rise of the stairs. In a house, this is the distance from the lower finished floor to the upper finished floor. For our activity, the total rise of the stairs will be 30″.

1. The first calculation will determine the number of risers in the stairs. The IRC code states the maximum riser height is 7 3/4″. A comfortable riser height for most people is 7″. To determine the number of risers, divide the total rise by 7.

   30″ ÷ 7 = 4.28 Round the answer to 4 whole risers

2. To determine the height of each riser, divide the total rise by the number of risers just calculated.

   30″ ÷ 4 = 7 1/2″ 7 1/2″ will be the unit rise for the layout

3. Calculate the tread width next. The sum of one riser and one tread should equal 17″–18″. A good tread width for this riser height would be 10″.

   10″ + 7 1/2″ = 17 1/2″ 10″ will be the unit run for the layout

4. The unit rise is then used to calculate the total run of the stairs. There is always one fewer tread than risers in the total run of the stairs. There will be three treads in this stair design, each 10″ wide.

   10″ × 3 = 30″ 30″ will be the total run

Stringer Layout

5. Set the 2×10 stringer on a set of sawhorses for the layout and cutting. Place the framing square on the edge of the 2×10, hold the tongue of the square on 7 1/2″, and hold the body of the square on 10″. Open the screw on a stair gauge and slide it onto the square. Slide one on the tongue of the square up to and touching the board and tighten the screw to secure the gauge in place. Do the same on the body of the square. See **Figure 1**.

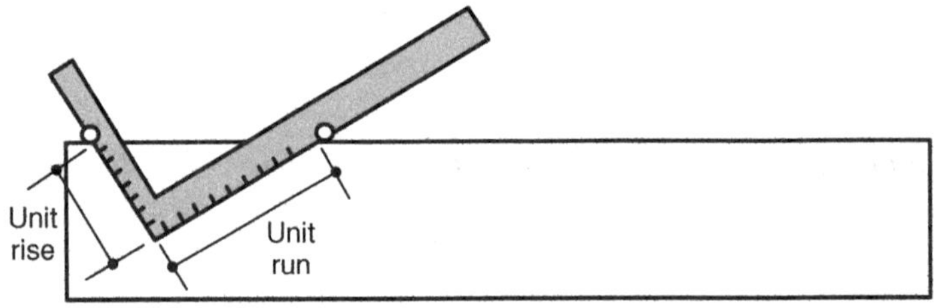

Goodheart-Willcox Publisher

Figure 1

> **NOTE**
> Once the stair gauges are secured to the square in the desired position, the square can be moved around, turned, or flipped to mark the layouts needed. The stair gauges will ensure the same angles will be drawn when they line up on the edge of the 2×10.

6. Start with the tongue of the square close to one end of the stringer, holding it tight to the edge of the board against the stair gauges. Draw a line along the outside of the square on both the tongue and the body. This will be the top step of the stairs.

7. Move the square along the board until the 7 1/2″ on the tongue lines up with the mark from the body in the previous step, **Figure 2**. Draw a line along the outside edge of the square again.

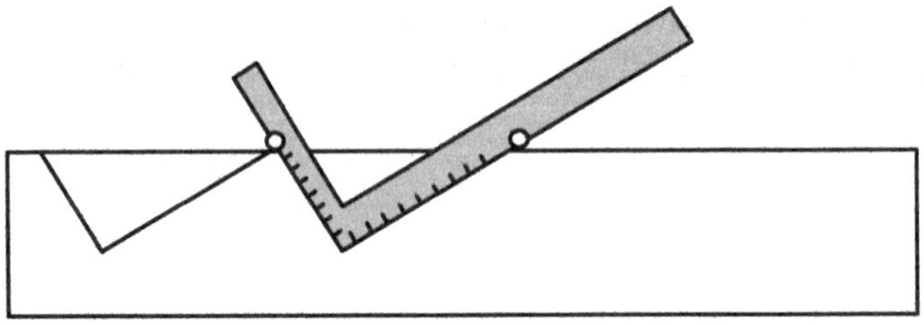

Goodheart-Willcox Publisher

Figure 2

8. Repeat the process again for the third tread. This will be the three treads required for the stair design.

9. Slide the square down the board one more time to help layout the bottom riser. Draw the lines along the outside of the square as before.

Name ______________________________

10. Pick the square up and without turning it or flipping it, move it to the other side of the 2×10. Align the body of the square with the last tread line, **Figure 3**, and draw a line along the square to the bottom of the stringer.

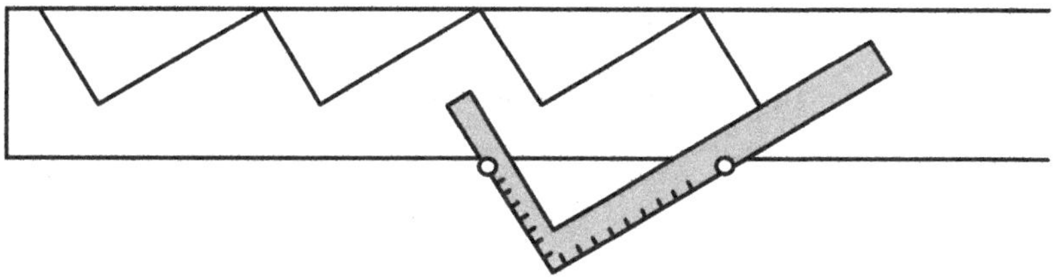

Goodheart-Willcox Publisher

Figure 3

11. Keeping the square on the bottom edge of the stringer, slide it to the top end of the stringer and align the tongue of the square with the first riser line. Draw a line from the riser line to the bottom of the stringer, **Figure 4**. This will be the top of the stringer fastened to the upper floor framing.

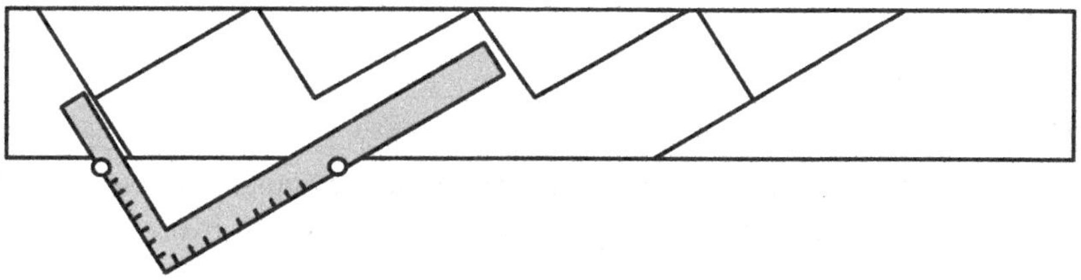

Goodheart-Willcox Publisher

Figure 4

12. The calculations done in the beginning steps do not allow for the thickness of the tread material being added on top of the layout lines. The bottom riser must be shortened by the thickness of one tread to ensure the first riser is the same as the others after the tread material is added. The 3/4″ OSB in the material list will be the tread material. Measure up from the bottom line on the stringer 3/4″ and draw a new line with the framing square parallel to the first. This will be the bottom of the stringer resting on the floor.

Cutting the Stringer

13. The stringer layout should match **Figure 5**. Each riser and tread line will be cut first to keep the board as long as possible so it can be held more safely. Then the top and bottom end cuts will be made to complete the stringer.

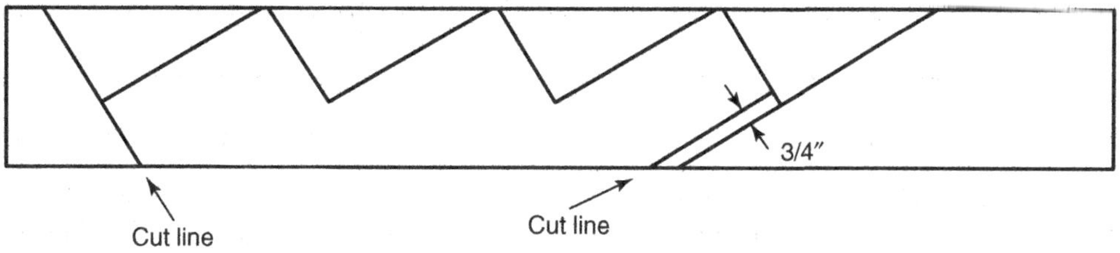

Goodheart-Willcox Publisher

Figure 5

14. Clamp or have a helper hold the 2×10 on the sawhorses so cutting can be done safely.
15. Cut each of the riser lines with the circular saw. Do not cut past the tread lines with this saw at the corner. If the saw cut is made too far into the 2×10, it will weaken the stringer.
16. When the corner is reached, turn off the saw and back the saw out after the blade has come to a complete stop.
17. Cut each of the tread lines stopping at the corner meeting the riser cut and follow the same procedure for removing the saw from the cut. The triangle piece left should stay in place.
18. Use the handsaw to cut from each side of the corners so the triangle pieces will fall out.

19. Cut the top and bottom layout marks. The final stringer should look like **Figure 6**.

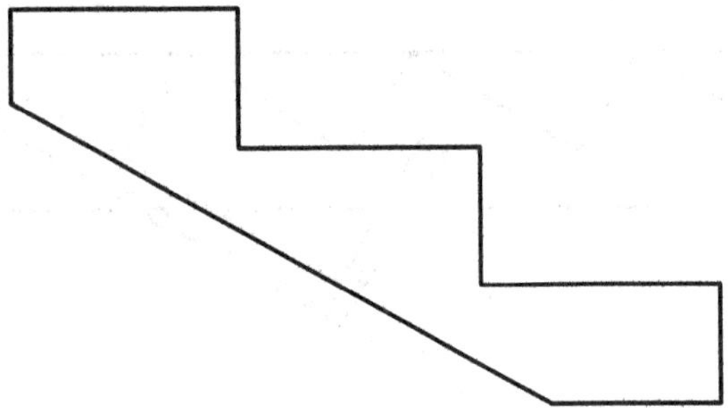

Goodheart-Willcox Publisher

Figure 6

20. Two stringers are needed to frame the stairs. Use this as a pattern to cut the second stringer by laying it on the remainder of the 2×10 and tracing each of the cuts to match. Cut the second stringer in the same manner as the first.

Building the Stair

Constructing the stairs will require some measurements. The practice stair being built will be 15 3/4″ wide. The treads will overhang the sides of the stringers on each side 7/8″. Therefore, the overall width of the stringers will be 14″.

21. Using the 2′ level, draw a plumb line from the floor up the wall of the corner unit to mark the first side of the stair. Slide the level up the wall while keeping it plumb so the line is at least 30″ tall. Measure to one side 14″ to locate the other side of the stair and draw another plumb line to the same height.
22. On the floor, use the framing square to draw a line out from and square with the wall at each of the stringer lines. Slide the square out, keeping it on the line and draw the line a bit more than 30″ from the wall.
23. On the wall of the corner unit, measure along one of the stringer lines up from the floor 30″ to locate the total rise of the stair.

NOTE

In a home, this would be the top of the upper finished floor that was used to do the calculations for the stair design. The floor framing is usually the top riser in the stair system. Because of this, the stringer is designed to intersect the framing one riser below the total rise mark. The top of the stringer is located by measuring down the height of one riser plus the thickness of one tread. Our riser is 7 1/2″ and the tread material is 3/4″ so measure down a total of 8 1/4″.

24. Using the level, draw a level line at this point between the two stringer lines. This is the top of stringer line.
25. On the floor of the corner unit, measure out from the wall the total run of the stair to locate where the end of the stringer will rest. On the lines previously made on the floor, measure out and mark at 30″. Draw a line between the two marks for a guide where to fasten the kicker. Refer to **Figure 7** for these measurements.

Name ______________________________

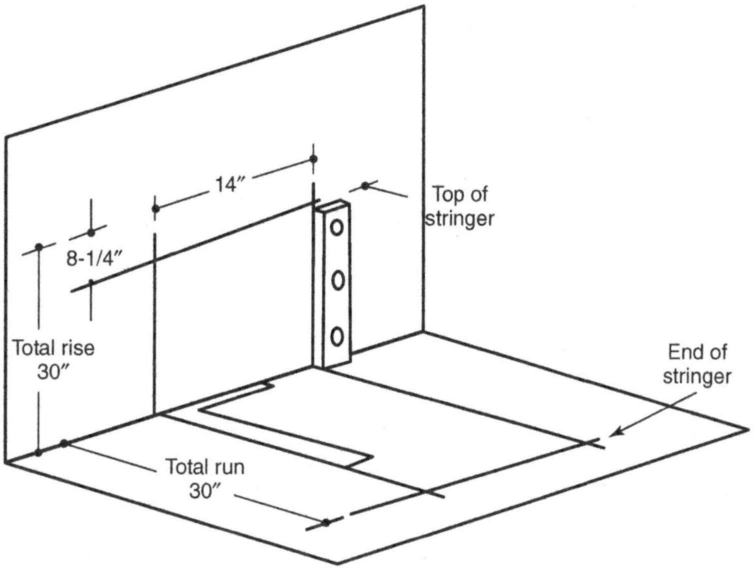

Figure 7

26. The top of the stringer will be fastened to the upper framing using a block between the stringers as a nailer. Cut an 11″ piece of 2×4 for this nailer. Measure and mark 1 1/2″ in from one of the stringer lines along the top of stringer line drawn in Step 24 to center this nailer. Fasten the block in place so the top of the block is at the top of stringer line and one end is at the 1 1/2″ mark. Some blocking may be needed in the wall to provide a solid backer to fasten to.

27. To provide solid support at the bottom of the stairs, a 2×4 kicker will be fastened to the floor. The stringer will be cut to fit over it for full support. Cut a 14″ piece of the 2×4 and fasten to the floor inside the line drawn 30″ from the wall and aligned with the lines at each end. The unit should look like **Figure 8**.

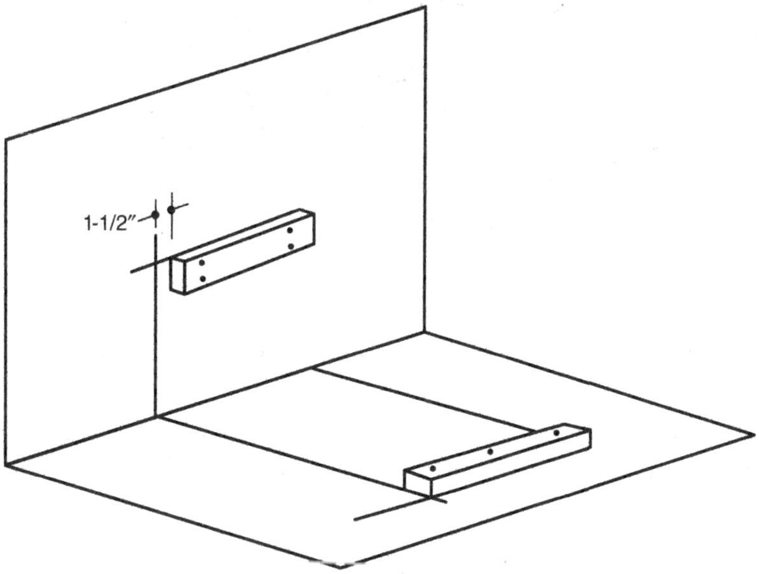

Figure 8

28. The stringers must be cut to fit over the kicker fastened to the floor. Measure up from the bottom of the stringer 1 1/2″ and draw a line. Measure in from the end of the stringer 3 1/2″ to layout the size of the 2×4 kicker as seen in **Figure 9**. Cut this out of the stringer. As was done before, do not cut past the corner with the circular saw so the wood is not weakened. Cut to the corner from each direction and finish the cut with the handsaw.

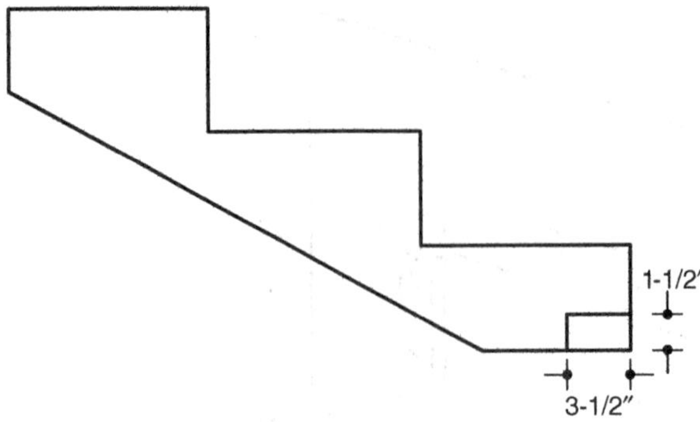

Goodheart-Willcox Publisher

Figure 9

29. Set the stringers in place and fasten them. The top tread surface of the stringer should be even with the top of the wall nailer and the bottom should fit over the 2×4 kicker on the floor. At the top, drive two screws through the stringer into the end of the nailer. At the bottom, drive a screw at an angle through the stringer into the top of the kicker, keeping the side of the stringer flush with the end of the kicker. Keep the screw back from the end of the stringer so it doesn't split the wood when driving it in. The project should resemble **Figure 10** at this point.

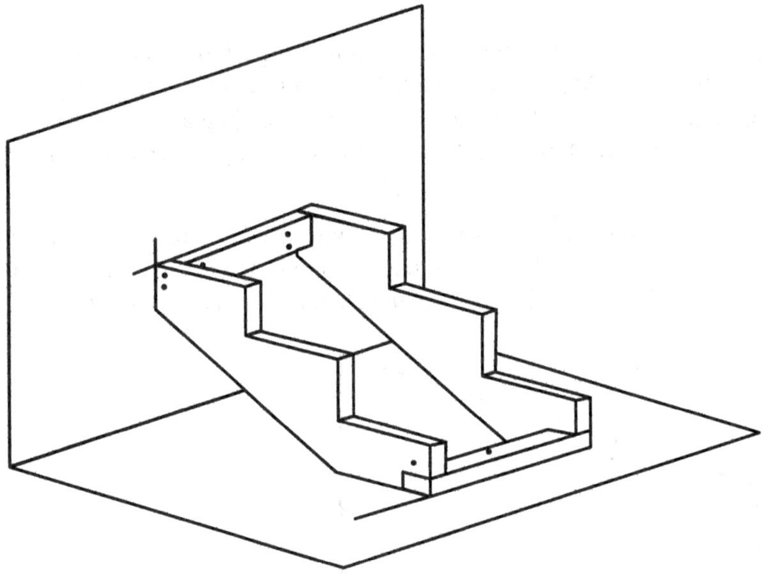

Goodheart-Willcox Publisher

Figure 10

30. The final step is to cut and secure the treads on the stringers. Cut three pieces for the treads from the 3/4″ OSB at 15 3/4″ each. Each tread should be 11″ wide and 15 3/4″ long.
31. Set the tread in place and line it up with a 7/8″ overhang on each side. Fasten in place using 1 5/8″ screws while keeping it tight against the back of the stringer. The finished product should look like **Figure 11**. Congratulations on your first stair build!

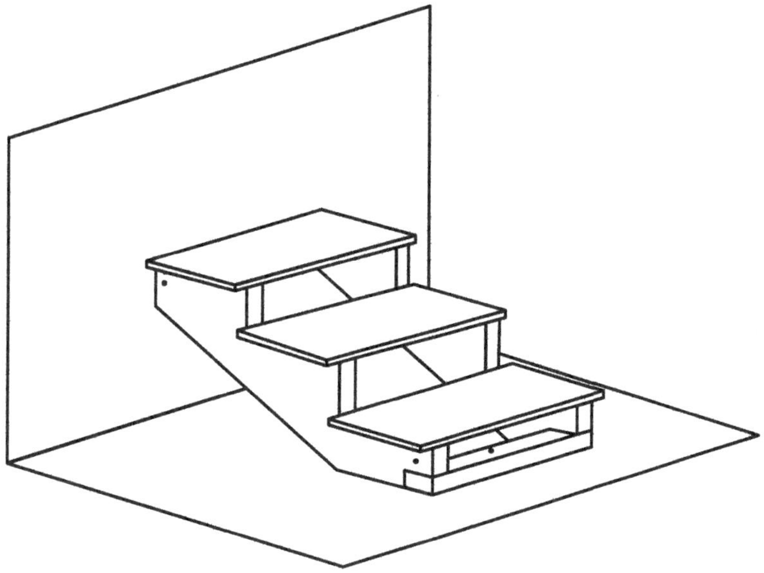

Goodheart-Willcox Publisher

Figure 11

Review and Analysis

1. Describe why 7″ is used as the first divisor in the first step of calculations to design the stairs.

2. Describe what would happen to the stringer if the cuts made in Step 17 were continued with the circular saw until the triangle piece fell out.

3. What is another name for the stair stringer?

Notes

Name ______ Date ______ Class ______

CHAPTER 9 Finish MEP Systems and Punch List

Activity 9-1: Chapter 9 Review

Answer the following questions after studying the chapter.

Lesson 9.1

1. _____ *True or False?* Small residential construction companies typically allow margin in their production schedule to send employees out for training on new technology. (9.1-1)

2. _____ *True or False?* Many types of construction management software allow workers to access documents from their smartphone. (9.1-1)

3. _____ Thermal imaging cameras are commonly used in residential construction to _____. (9.1-1)
   A. measure distances
   B. reduce muscle strain and fatigue
   C. inspect inside drainpipes
   D. evaluate HVAC systems

Lesson 9.2

1. _____ *True or False?* Renovating is more transformative than remodeling. (9.2-1)

2. _____ *True or False?* New construction generally takes longer than remodeling and renovating projects. (9.2-1)

3. _____ *True or False?* The use of asbestos in residential construction was discontinued in 1989. (9.2-1)

4. _____ Mold in homes is associated with _____. (9.2-1)
   A. respiratory issues
   B. colon cancer
   C. learning disabilities
   D. seizures

Lesson 9.3

1. _____ *True or False?* Shower units are installed after interior surfaces and trims are complete. (9.3-1)

2. _____ One task of finishing out the HVAC system is _____. (9.3-1)
   A. installing ductwork throughout the house
   B. adjusting dampers to ensure proper airflow
   C. placing heating elements in heated floors
   D. deciding which type of furnace will be installed

3. _____ Who is responsible for assigning punch list tasks? (9.3-1)
   A. Homeowner
   B. Subcontractor
   C. General contractor
   D. Architect

4. _____ *True or False?* The certificate of occupancy must be issued before the final building code inspection takes place. (9.3-2)

Lesson 9.4

1. _____ *True or False?* The terminals on electrical switches are usually screws. (9.4-1)

2. _____ When installing an electrical receptacle, the neutral wire should be attached to the _____ terminal. (9.4-1)
   A. silver
   B. green
   C. copper
   D. red

3. _____ What type of plumbing supply is used to seal a P-trap connection? (9.4-1)
A. O-ring
B. Slip-joint washer
C. Pipe compound
D. Plumber's putty

4. _____ The proper way to use Teflon® tape is to wrap it around the _____. (9.4-1)
A. female threads before connecting the pipes
B. male threads before connecting the pipes
C. entire joint after the pipes have been connected
D. entire pipe, from one end to the other

5. _____ Which is an example of hardscaping? (9.4-1)
A. Outdoor fireplace
B. Outdoor floodlight
C. Seeded lawn
D. Sodded lawn

6. _____ *True or False?* A wing nut is used to connect two or more wires together. (9.4-1)

Lesson 9.5

1. _____ A multimeter can be used to measure _____. (9.5-1)
A. only voltage and amperage
B. only voltage and resistance
C. only amperage and resistance
D. voltage, amperage, and resistance

2. _____ What tool shears off small bits of sheet metal with a mechanized punch that has a similar action to a sewing machine? (9.5-1)
A. Aviation snips
B. Powered metal shears
C. Nibbler
D. Router

3. _____ *True or False?* Flexible copper tubing connections are commonly made with flared fittings. (9.5-1)

4. _____ *True or False?* Ratchets and sockets are tools used on fasteners with a square head. (9.5-1)

5. _____ *True or False?* Impact drivers can be powered by air. (9.5-1)

Lesson 9.6

1. _____ *True or False?* An accident victim can receive additional injury from unskilled treatment. (9.6-1)

2. _____ The employer is responsible to ensure that a person who is certified in first-aid training is available at the worksite if a medical professional is not able to respond to an emergency within _____. (9.6-1)
A. 3 to 4 minutes
B. 10 to 12 minutes
C. 20 to 30 minutes
D. 45 to 60 minutes

3. _____ What is the proper way to dispose of a utility knife blade that is contaminated with blood? (9.6-1)
A. Throw the blade into the trash.
B. Retract the blade into the utility knife and throw the entire thing away.
C. Place the blade in a designated sharps container.
D. Wrap the blade in duct tape and then place it in a sharps container.

4. _____ *True or False?* Administering CPR immediately after a person's heart stops beating can double or triple their chance of survival. (9.6-1)

5. _____ *True or False?* Some accidents can cause unsafe circumstances for uninvolved employees after the event has occurred. (9.6-1)

6. _____ Which statement about emergency action plans (EAPs) is true? (9.6-2)
A. A properly designed EAP can apply to any jobsite without making any changes.
B. The employer must ensure individuals are trained in the elements of the EAP.
C. Simply having an EAP in place makes workers safe on the job.
D. An EAP should primarily focus on what to do in the event of a fire on a jobsite.

Name ____

Lesson 9.7

1. _____ Estimating means to _____. (9.7-1)
   A. take an accurate measurement of something
   B. determine accurate quantities of something
   C. prepare an invoice for a client
   D. pay an invoice issued by a vendor

2. _____ How much concrete should you order if you need to pour a patio slab that is 14′ × 30′ × 4″? (9.7-1)
   A. 5 yd^3
   B. 6 yd^3
   C. 140 yd^3
   D. 150 yd^3

3. _____ To determine the number of joists needed for 24″ OC spacing, you _____. (9.7-1)
   A. divide the length of the wall by 3/4
   B. divide the length of the wall by 1/2
   C. multiply the length of the wall by 3/4
   D. multiply the length of the wall by 1/2

4. _____ *True or False?* When estimating the subflooring, you calculate the area by multiplying the overall length and width, and then subtracting the major areas that are not to be covered. (9.7-1)

5. _____ When estimating subflooring that will be made with sheet material, after determining the net area, you _____ and round out to the next whole number. (9.7-1)
   A. multiply by 24
   B. multiply by 32
   C. divide by 24
   D. divide by 32

6. _____ How many 2″ × 4″ × 16′ pieces of lumber will you need to order if you are using a double top plate and the total wall and partition length is 360′? (9.7-1)
   A. 57
   B. 68
   C. 75
   D. 82

7. _____ When estimating studs that are spaced 16″ OC, you should _____. (9.7-1)
   A. multiply the total length (in feet) by 3/4
   B. multiply the total length (in yards) by 3/4
   C. divide the total length (in feet) by 3/4
   D. divide the total length (in yards) by 3/4

8. _____ How many sheets of 4′ × 8′ wall sheathing will be needed if the net area is 1800 sq ft? (9.7-1)
   A. 43
   B. 57
   C. 66
   D. 75

9. _____ When estimating the number of rafters required for a plain gable roof with 24″ OC spacing, you must multiply the lengthy of the building by _____. (9.7-1)
   A. 1/2, then round up and add one
   B. 1/2, then round up and subtract one
   C. 3/4, then round up and add one
   D. 3/4, then round up and subtract one

10. _____ If the roof slope is 3-in-12, and the ground area is 1400′, how many squares should you order if asphalt shingles are being used? (9.7-1)
    A. 15
    B. 16
    C. 17
    D. 18

11. _____ If the total perimeter of a home is 640 sq ft, the ceilings are 10′ high, and the total area of windows and doors is 150 sq ft, what is the net area that needs to be insulated? (9.7-1)
    A. 4970 sq ft
    B. 5467 sq ft
    C. 6250 sq ft
    D. 6875 sq ft

12. _____ *True or False?* Estimates for solid paneling are based on its dressed size. (9.7-1)

13. _____ *True or False?* There are typically 36 bd ft in a standard bundle of wood strip flooring. (9.7-1)

14. _____ If you need to order 12″ × 12″ tiles for a rectangular room that is 18′ long and 15′ wide, and the chart indicates that 7% should be added for waste, how many tiles will you need? (9.7-1)
A. 289
B. 297
C. 385
D. 396

15. _____ *True or False?* When estimating porcelain tile, you should add an additional 10% to compensate for breakage, cuts, and extras for future repairs. (9.7-1)

Lesson 9.8

1. _____ The process of determining the quantities of specific materials that will be required for a given project is called the _____. (9.8-1)
A. material order
B. labor estimate
C. change order
D. material takeoff

2. _____ Which type of information is commonly found in a request for information? (9.8-1)
A. References from former clients
B. Appeal for a response
C. Estimate for materials and labor
D. Proposal for a second project

3. _____ If the client wants to alter the original scope of work, they must sign a _____, which outlines the requested modifications and related information. (9.8-2)
A. request for information
B. material takeoff
C. change order
D. punch list

Lesson 9.9

1. _____ Combustion sources such as gas stoves and fireplaces can release _____ and other harmful pollutants into the home if not properly vented or maintained. (9.9-1)
A. carbon monoxide
B. nitrous oxide
C. dihydrogen oxide
D. dinitrogen pentoxide

2. _____ Long-term exposure to radon is linked to which medical condition? (9.9-1)
A. Lung disease
B. Alzheimer's disease
C. Heart disease
D. Addison's disease

3. _____ Testing indicates that indoor air should be exchanged with fresh outdoor air every _____. (9.9-2)
A. 30 to 60 minutes
B. 2 to 3 hours
C. 10 to 12 hours
D. 2 to 3 days

4. _____ *True or False?* If the homeowner fails to operate and maintain ventilation systems, the air quality will begin to deteriorate quickly. (9.9-2)

Name ______________________________ Date ______________ Class __________

Activity 9-2: Project—Installing Electrical Fixtures

Learning Outcomes

After completing this lab activity, you will be able to:

- Install a duplex outlet.
- Install a switch and light fixture with future safety in mind.

Introduction

With the electrical rough-in complete, it is now time to install electrical devices so the electrical circuit is functional. Power will first come to the duplex outlet. It will be connected to the duplex outlet in such a way that it will power the outlet and will provide power to the light as well. All fixtures will have a ground wire in place, if a grounding location is provided on the device. All work will be done with the model unplugged. No power will be engaged unless checked by the instructor.

Text Reference

Chapter 9, *Finish MEP Systems and Punch List*

Equipment

The following tools and equipment are needed to perform these activities:

- #2 Phillips screwdriver
- #2 standard screwdriver
- Duplex outlet
- Duplex outlet cover
- Light bulb
- Linesman's pliers
- Needle-nosed pliers
- Plug-in circuit tester
- Porcelain light fixture
- Safety glasses
- Single pole light switch with ground
- Switch cover
- Wire nuts, 2-wire (1) and 3-wire (2)
- Wire stripper

Safety Review

Follow all safety procedures specified by your instructor and specified in any appropriate manufacturer's information. The following are some additional safety-related items to keep in mind:

- Wear safety glasses at all times.
- Before installing any electrical fixture or making adjustments on any electrical item, make sure all the power is turned off at the circuit breaker or unplugged.
- All work done on the construction model will be with the model unplugged.
- Consider all electrical wires as HOT, even if the power is turned off.

Procedure

All electrical devices should have a place for the hot wire, neutral wire, and usually a ground wire. Hot (black) wires get attached to the *brass* or dark-colored screws, Neutral (white) wires get attached to the nickel or light-colored screws, and ground (green or bare copper) wires attach to the green hexagonal screws. Each box should have wire leads of 6″ that can be pulled out to attach devices such as light switches, outlets, and lights or pigtails. A pigtail refers to when a pair of wires must be connected but also needs to be connected to the device. A short piece of wire is connected to the two wires with a wire nut and then connected to the device.

Follow the wiring diagram shown in **Figure 1** while working through this activity. It shows where each wire is to be connected to each device along the circuit.

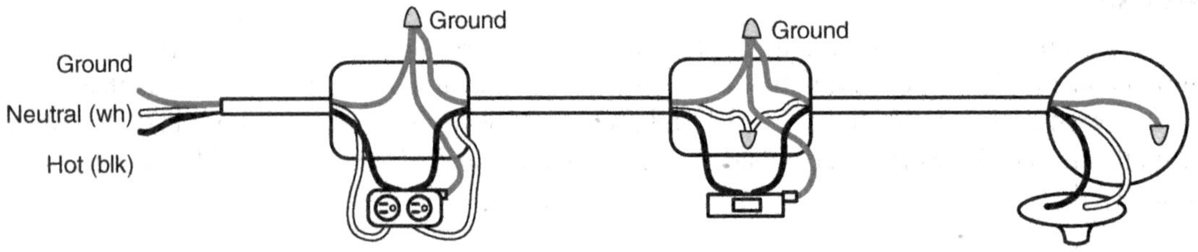

Goodheart-Willcox Publisher

Figure 1

1. Start at the outlet box by pulling the wires out of the box and straightening them out to work on. Push the black wires to the right side of the box and the white wires to the left. That is the side they will be on when they are attached to the outlet.
2. A pigtail needs to be added to the ground wires first because the ground wire needs to be continuous through the circuit but also needs to be attached to the outlet. Cut an 8″ piece of ground wire from another piece of NM cable. Hold all three wires together and twist the ends together using the lineman pliers then place a 3-wire wire nut on the ends and twist it in a clockwise direction until tight to hold them together.
3. Fold the ground wires with the wire nut back into the box out of the way with just the pigtail coming out to the outlet.
4. Each of these wires will be attached to the outlet device with the screws on the sides of the device. This will be accomplished with a loop on each to go around the screw before tightening. Use needle-nosed pliers or strippers to make a loop on each, **Figure 2**.

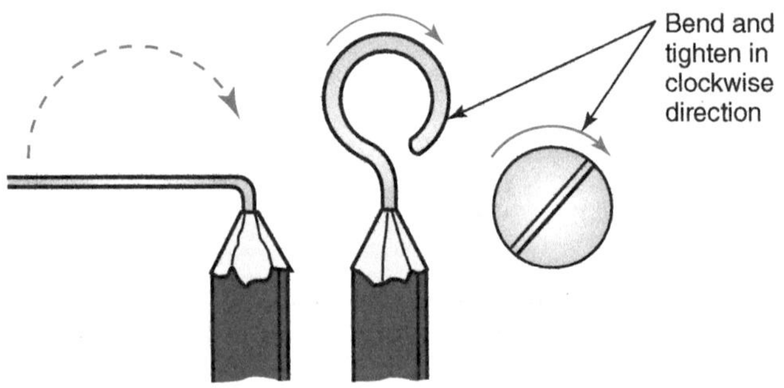

Goodheart-Willcox Publisher

Figure 2

NOTE

Near the middle of the jaws of the wire strippers along each side are two small holes through the blades. This hole can be used to create the loop needed to attach the wire to the screws of the device. Put the wire in the hole with just the very end sticking past the other side. Turn the stripper and bend the wire around the side of the stripper until the end is parallel to the wire making a J shape. This loop will fit neatly over the device screw to make the connections.

Name ____

5. Connect the wires to the duplex outlet. Connect Hot (Black) wires to the side of the duplex outlet with the brass screws, Neutral (white) wires to the side with the nickel (silver) screws, and the ground wire to the green screw. Be sure to put the loop so it goes around the screw in a clockwise direction. This will pull the loop in tight to the screw when it is tightened.
6. Fold the extra wire back into the box and attach the outlet to the box using the screws on the outlet ears.
7. At the switch box, the feed or hot side of the circuit in the black wires must be wired to the switch to ensure there is no power to the light unless the switch is turned on. If the black wires were connected here, the light box would have power to it all the time which creates a hazard for someone working on the light fixture.
8. The return or neutral wires will be connected here to complete the circuit. To do this, connect the two neutral (white) wires with a 2-wire nut. Twist them together first with the lineman pliers to get a good connection inside the wire nut. Fold them back into the box after connected.
9. Attach the black wire coming from the outlet box to the bottom screw of the light switch. Make sure the on/off is upright and not upside down when you attach your wires. When the switch is right side up, it says "ON." If it is upside down, it says "NO."
10. Attach the black wire coming from the light to the top screw of the light switch.
11. A pigtail is needed on the ground wire to make the connection to the ground screw on the switch. Cut an 8″ piece of ground wire from another piece of NM cable and twist all three wires together with the lineman pliers. Make a loop on the pigtail end and attach it to the green screw on the switch.
12. Roll the extra wire into the back of the box and attach the light switch to the box.

NOTE

If the switch does not have a ground screw, connect the two ground wires with a 2-wire nut and fold them into the back of the box before attaching the switch.

13. Pull the wires out of the octagonal box for the light, form the termination loops, and attach the wires to your light, **Figure 3**. If there is no green ground screw on the light fixture, cap off the ground wire with a wire nut.

Goodheart-Willcox Publisher

Figure 3

14. Fasten your light to the round light box.
15. Attach the light switch cover and duplex outlet cover to each of those devices using the screws provided in the cover plate package.
16. Put the circuit tester in the duplex outlet and a light bulb in the light fixture. Have the teacher plug in your model so you can check if wiring is correct and verify that the light switch/light works. See **Figure 4**.

Goodheart-Willcox Publisher

Figure 4

17. If something does not work, unplug the extension cord, troubleshoot and correct the wiring if needed and retest to verify.

Review and Analysis

1. When attaching the loop of wire to the device screws, why is important to have the loop going around the screw in a clockwise direction?

____

____

____

2. How would you troubleshoot a light not working properly?

____

____

____